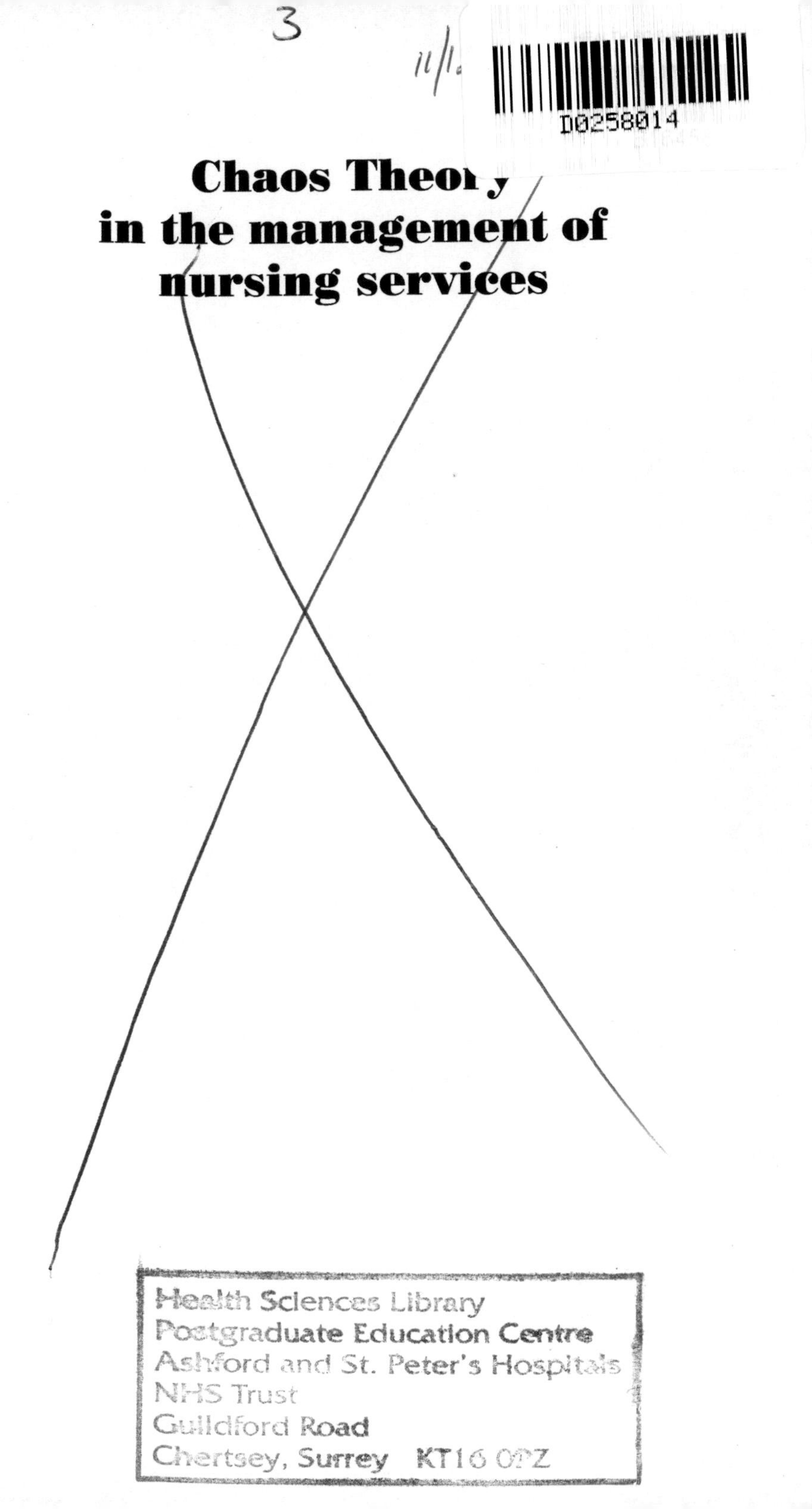

Chaos Theory in the management of nursing services

Chaos Theory in the management of nursing services

Carol A Haigh

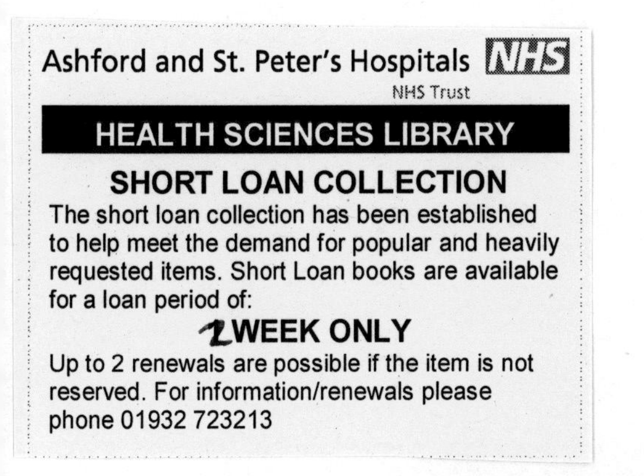

APS Publishing
The Old School, Tollard Royal, Salisbury, Wiltshire, SP5 5PW
www.apspublishing.co.uk

British Library Cataloguing in Publication Data
A catalogue record for this book is available from the British Library

ISBN 1 9038770 4 0

Printed in the United Kingdom by Biddles Ltd., Guildford and King's Lynn

Contents

Acknowledgements

When I was doing my Post Registration Diploma in Nursing in the 1980s, I had a fantastic lecturer whose name was Josie Evans, and who first suggested to me that I should try to have my work published—the rest, as they say, being history, I feel I owe her a lot. Since then, and certainly while this book was in production, there have been so many people to acknowledge that, unfortunately, there is not space to mention them all here. However, there are certain individuals whose support has been above and beyond the call of duty. Bryan, who has always been there to love and encourage me, even when I was absent for so many long periods in my study that he could only recognise me by the wedding photos. My mother, who put up with long conversations about Chaos Theory when she would rather have been watching 'Fraiser' on TV. My sister, Jacqueline, who bravely ploughed through the first draft of this work and my other sister, Kay, who didn't but provided retail therapy whenever she could. My two good buddies, Stephen Bailey and Neil Jones, who supplied endless lunches to sustain me and bizarre humour to keep me grounded. My best mate Carol Tanner and, finally, Flix, Zesta and Spooky, my three adorable cats. I'm not sure that a book on Chaos Theory is really what any of them would have wanted, but this book is dedicated to all of them with my love and gratitude.

Carol Haigh
October 2002

Introduction

Although this book is not strictly speaking about physics or even much about Chaos Theory, both these elements come into it. To understand where Chaos has come from, it is important to have some insight into how the science of physics developed. The purpose of this brief, introductory segment is to provide a preamble to Chaos Theory by giving a short and necessarily superficial review of the history of physics. This is because, in order to contextualise the concept of Chaos Theory, it is necessary to furnish a brief overview of the development of physics as a discipline. This will assist in identifying and clarifying the links that can be seen to exist between current thinking in terms of Chaos Theory and the elements of Chaos within both classical and quantum physics. Don't panic, this is in the nature of an interesting foundation to the ideas that follow within the rest of the book and you can skip it if you want to; it won't affect your understanding of the rest of the work, but that understanding will be a little less rich than if you persevere. This synopsis will be supplemented by an in-depth consideration of the individual elements of Chaos Theory in the following chapters. In this manner, the components of Chaos will be clearly articulated as a precursor to the practical application of Chaos Theory to a specific nursing system that forms the major focus of this book.

From Classical to Quantum to Chaos: The evolution of physics

Although the origins of physics can be traced back as far as 450 BCE (before common era) when Leucippus of Miletus speculated upon the concept of the atom, it can be argued that the real birth of modern physics took place in the late fifteenth and sixteenth centuries. Observations, such as that of

Columbus in 1492, who noted that a magnetic compass points in different directions at different longitudes, and Galileo Galilei who produced an erudite description of how bodies fall under gravity (even though the law of universal gravitation would not be formally articulated for another 86 years), produced insights into the natural world that changed perceptions and acted as catalysts for further study (Barnes-Svarney, 1995; Stewart, 1997).

The foundations of what is today referred to as classical physics were laid in the years from 1666, when Newton published his invention of Calculus and developed his Law of Universal Gravitation, to the mid nineteenth century with Kelvin's proposal of absolute zero in 1848 and Maxwell's discovery of electromagnetism in 1873 (Barnes-Svarney, 1995). With the establishment of the essential elements of classical physics, namely Newtonian mechanics, thermodynamics, acoustics, optics and electricity, the discipline of physics was ready to extend its conceptual boundaries. The impetus for this would come not from a theorist, but from a classical physicist, Max Planck.

In the years between 1895 and 1900, Planck was working upon the problem of blackbody radiation. A black body is an idealised example of a 'perfect' absorber of radiation. A simple example of a black body is a hollow sphere or tube with sealed ends and a small pinprick hole in one side. Any radiation, such as light, that enters the hole will be trapped within the sphere until it is absorbed. It is highly unlikely that it would manage to escape through the hole by which it entered. Planck was attempting to resolve the dichotomy between the two approximate equations that described blackbody radiation at the long wave length and short wave length end of the blackbody spectrum. The difficulty was that both laws, the Rayleigh-Jeans Law of Long Wave Radiation and Wien's Law of Short Wave Radiation, did not describe the whole spectrum of blackbody radiation. Therefore, the events at opposite ends of a continuum were understood, but no theory existed to explain the whole phenomenon—a situation analogous to modern medicine understanding the nature of conception and childbirth, but having no insight into the development processes that occur in between the two events. Planck was sure that the answer lay in the connection between thermodynamics and electrodynamics, but it was only when he combined both laws into one simple mathematical formula, combined with a little mathematical 'tweaking', that he was able to produce a formula to describe the whole spectrum.

What made this a turning point in the history of physics, is that Planck's solution to the problem had no experimental or physical basis. Instead of building the answer up from physical assumptions, the answer had just 'appeared'. The effort expended by Planck in subsequent years to synthesise a physical explanation of the mathematical equation, particularly his incorporation of Boltzmann's statistical version of the Second Law of Thermodynamics (natural processes always move towards an increase in disorder, or entropy always increases) which was a way of expressing entropy mathematically, led to the development of quantum theory (Gribbin, 1984). By 1921, Einstein received the Nobel Prize for his application of quantum theory to describe photoelectric effects. He had done this by abandoning classical physics and accepting Planck's equations as physically meaningful without experimental 'proof'.

This abandonment of the laws of classical physics can be argued to be the greatest contributory factor in the subsequent development of physics. The description of the basic structure of the atom, produced by Rutherford and Bohr in 1913, only appears stable if quantum rules are used to describe the behaviour of electrons. This paradigm shift in thinking about physics heralded the development of quantum physics and quantum theory in particular. Schrödinger, building upon the work of de Broglie, developed the wave version of quantum mechanics in 1925. Heisenberg's Uncertainty Principle, which states that it is impossible to identify where a particle is *and* how fast it is travelling, postulating that the act of observing disturbs the observed, was developed as an effort to resolve the conflict of concepts between the particle hypothesis of Bohr and the wave theory of Schrödinger. Concepts such as 'position' and 'velocity' have different meanings at the atomic level and the uncertainty principle addresses this (Gribbin, 1984)

The notion that attempts to measure position and velocity of a particle will affect the behaviour of that particle is the first faint hint of Chaos Theory in the quantum world. Schrödinger made an attempt to clarify the concept with his famous ' cat in the box' thought experiment, although his intention was less to clarify and more to highlight the foolishness of Bohr's complex and conceptual measurement hypothesis, sometimes referred to as the Copenhagen interpretation of quantum mechanics. Bohr suggested that the quantum world exists in a superposition of states (a sort of potential of all possible outcomes state) until it is measured. To put it crudely, the process of measurement then 'fixes' the item of interest into that measured state. Although the Copenhagen interpretation has elements of the bizarre, it is still pre-eminent today (Gribbin, 2002). Schrödinger, in order to highlight the absurdities inherent in this interpretation, postulated that if a live cat is placed in a box with a vial of poison that will shatter when a Geiger counter detects the decay of some radioactive material (also in the box), then an external observer will not know the fate of the cat until they lift the box lid and look inside. This act of observing will 'fix' the cat in either a dead state or a live state (Stewart, 1997). However, this clever thought experiment slightly backfired upon Schrödinger when it became apparent that his work was being used to support the notion of a probability wave rather than to emphasise its inherent absurdity. Disillusioned, Schrödinger is reported to have said of his thought experiment, 'I don't like it and I wish I'd never and anything to do with it' (Gribbin, 2002).

The suggestion that the universe does not run rigidly in accordance with the laws of classical physics was contextualised within the uncertainty that was inherent in quantum physics. Hawking (1987) noted that Heisenberg's uncertainty principle undermined the notion of a completely deterministic universe, in that this reformulation of classical physics opened the way for unpredictability and randomness into the universe. Although earlier Feynman challenged this view to some extent by suggesting that quantum mechanics should not be viewed as an escape from a completely deterministic universe, he did go on to acknowledge that classical laws were flawed, in that tiny errors, small gaps in understanding could be amplified by the interactions of dynamic systems until they reach large scale conclusions (Feynman, 1965).

Hunt and Yorke (1993) argue that James Clerk Maxwell who explained electromagnetism and published work on the kinetic movement of gases in 1860 was the first person to understand this aspect of Chaos. Maxwell's clearest expression of the concept was delivered in an essay at Cambridge University in 1873 and bears replication here,

> *'When the state of things is such that an infinitely small variation of the present state will alter by only an infinitely small quantity, the state at some future time, the condition of the system... is said to be stable; but when an infinitely small variation in the present state may bring about a finite difference in the state of the system in a finite time, the condition of the system is said to be unstable.'*
>
> (Cited by Hunt and York, 1993: 3)

Once again, the faintest hint of Chaos Theory thinking can be detected in Maxwell's understanding that a complex system will show sensitive dependence upon initial conditions. This is one of the fundamental principles of Chaos Theory, and will be fully explored in the next chapter.

1

Introducing Chaos

This is not a book about Chaos as a science. There are many such books available and they are listed in the suggested reading section at the end of this book; many of them will also be found in the reference list. This is a book about how to use elements of Chaos Theory in a pragmatic way that will contribute to effective management of nursing care. Nursing care has become more complex, and resources in the shape of time and personnel have become increasingly scarce. As some elements of care fragment into specialist services, there is a need for practitioners to become familiar with concepts, such as business planning and objective time management. There are many time-limited methods that facilitate this type of planning, but few that can help practitioners to forecast the fate of the service they deliver. This book is all about using components of Chaos Theory to do just that. Chaos Theory is perceived as complex in the extreme, and indeed, it is. However, modern nurses are familiar with the ideas of 'process' since we incorporate the notion into our assessment of patient needs, and also in the writing of audit tools. So, the focus of this book will be to break one small element of Chaos into manageable parts that will allow us to grasp the process of applying chaotic thinking to predict service longevity. Therefore, this chapter will concentrate less upon the development of Chaos Theory as a practical science and more upon attempting to make the fundamental elements of Chaos accessible. This means that there has been a degree of selection regarding the issues that are presented. The overall aim is to allow the reader to become familiar with the fundamentals of Chaos only in as far as they will impact upon the identification of chaotic elements at a later stage.

Since the beginning of time, human beings have been trying to apply order to the natural world. From the ancients, who saw pictures in the stars and gave us the constellations, to modern children, who see meaningful shapes in clouds, the need to impose order on the apparently disordered could almost be argued to be a fundamental human urge. This requirement has given us a systematic understanding of how our world works. The word 'systematic' is used in two ways here. Firstly, our understanding has developed via logical and methodical investigation, with ideas and research being developed in an orderly manner in much the same way that we may 'systematically' examine a patient. Secondly, these investigations have given us insights into the various 'systems' that make up our world. Thus, in health care, we speak of digestive systems, reproductive systems, and respiratory systems. We can investigate these systems and draw up rules or boundaries that described the way they work and how they are likely to behave in certain circumstances.

However, some phenomena do not comfortably fit within this ordered or deterministic view of the Universe, patterns of teenage pregnancies, cardiac arrhythmias or the spread of an influenza epidemic, for example. For a long time, it was assumed that there was no detectable pattern or format to such phenomena but, eventually, it became apparent that these processes are not truly random. They just look as though they are when they are examined at a superficial level. The collective term for processes of this sort is *Chaos*. Chaos helps us to make sense of the world we live in and allows us to understand and predict how complex systems will develop. The person who is credited with the 'discovery' of Chaos is Edward Lorenz.

Lorenz, a meteorologist at the Massachusetts Institute of Technology, was studying weather patterns. He had set up his computer to model weather predictions with twelve variable patterns. By manipulating the patterns, Lorenz could change the weather model that his computer would produce. One day, when reviewing a particular sequence, Lorenz took what he thought was a short cut and began running the sequence in the middle instead of at the beginning; this was the first step towards encountering Chaos. The second step was when, instead of typing his data with its normal six decimal places, he only input three decimals in order to fit all of the numbers onto one line. Lorenz then went to get a cup of coffee and when he returned, he found a very strange thing had occurred. Instead of producing the familiar pattern that Lorenz expected, the computer had produced a totally different weather sequence. The two apparently trifling changes had been enough to alter the sequence dramatically.

What Lorenz had discovered was that the smallest change in the initial circumstances of a dynamical system will drastically affect the long-term behaviour of that system. This is now referred to as sensitive dependence upon initial conditions—in other words, Chaos. Sensitive dependence upon initial conditions means that a system will be affected by the conditions in place at the beginning. The popular expression of this is the 'butterfly effect'. Most people are familiar with the theory that if a butterfly flaps its wings today, the tiny changes in air pressure will eventually lead to a hurricane in China at some future point. A similar analogy is suggested by Ruelle (cited by Sardar and Abrams, 1999). Ruelle suggests that if a little devil wanted to upset your life, all they would have to do is alter a single electron in the atmosphere. At first, you would not notice anything, but the structure in the turbulence of the air has been affected. After about two weeks, the change has grown enough to affect weather systems and you are caught in a downpour on the way to an important interview. If it sounds as if I am labouring this point, I am. Sensitive dependence upon initial conditions is one of the first (probably the most important) concepts you need to grasp to make sense of what follows.

Both of the above examples demonstrate sensitive dependence upon initial conditions. This is a fundamental concept of Chaos, and an important one to comprehend. The butterfly flaps its wings; the devil moves an electron, thereby making a change in the initial condition of our weather systems. The sensitive dependence of the system means that these tiny changes end up as big effects later. Examples of sensitive dependence upon initial conditions are not exclusive to weather. Stopping to answer the ward telephone might mean that you are not the one to find a collapsed patient; without your particular expertise, resuscitation may be unsuccessful—sensitive dependence upon initial conditions.

Stewart (1997) argues that the butterfly scenario, which is without doubt the best known example of sensitive dependence upon initial conditions, is a simplified example. He contends that the best that the butterfly could hope to achieve would be that if a hurricane was going to occur anyway, it might occur a little earlier or later than it would otherwise. The 'little devil' scenario complements the 'butterfly' effect because it emphasises strongly that the initial change is minuscule in the extreme, but can eventually be translated into a large effect. The point that Stewart makes is true of our mischievous devil also; moving that electron will not create a weather system from nothing. It might only make the difference between being caught in a light shower or a full-blown thunderstorm. Likewise, with the example I gave on resuscitation, the patients might have been going to die anyway; they simply die earlier than they would have.

The important part of Lorenz's discovery was that the mathematics that allowed Chaos to be plotted and replicated had been found. The fundamental components of chaotic systems could be identified and the effect of parameter manipulation could be tested. By altering his data in an apparently unimportant manner, Lorenz had brought about a dramatic change. For the first time, apparently disordered systems were shown to exist with an underlying 'order' beneath the disorder. The elements of Chaos had been identified and, given sufficient system parameter information, could be predicted and represented graphically.

The elements of Chaos

The following is a technically accepted definition of Chaos that has been suggested:

> *'Processes that appear to proceed according to chance, even though their behaviour is in fact determined by precise laws.'*
>
> (Lorenz, 1993: 4)

At first glance, this definition, with its focus upon natural laws, appears redolent of the classical physics view of a deterministic universe, the notion that everything that happens is in accordance with natural laws. However, upon closer inspection, this interpretation is a lot more complex. It must be emphasised that what Lorenz refers to as 'processes' in the definition cited above, most other authors, notably Gleick (1987), Stewart (1997) and Kosko (1994), refer to as systems. Chaos is seen to be merely one outcome of any system's evolution. The elements of Chaos can be identified as:

- a dynamical system
- identified holon parameters
- a specified equilibrium state
- a specified state attractor.

Each of these elements contributes towards chaotic outcomes in system development and, as such, deserves explanation at this point.

Dynamical systems

Chaos, in any system, depends upon that system being dynamical. A dynamical system is one that changes with time. Kosko (1994) argues that, in theory, *everything* is a dynamical system in the sense that all systems begin with an initial condition and traverse through transient states to an equilibrium state. The initial condition of any dynamical state will be that of maximum potential energy. The end equilibrium state is typified by minimum potential energy. The initial state potential is transformed into kinetic energy. Physical objects may possess kinetic energy (the energy of motion) or potential energy (the objects store energy because of their position or consideration). For example, when you lift a patient using a mechanical hoist, the hoist uses kinetic energy to lift the patient against the force of gravity. Once the patient is raised up, the potential energy is in the patient's weight because if the patient is released suddenly the gravitational pull of the Earth will put him/her into motion. In a dynamical system, as the system moves from its starting point of maximum energy, through its intervening transient states to equilibrium, the systems original energy is released as kinetic energy.

To understand the concept of Chaos as we plan to apply it, it is important that we are completely clear about what we refer to as a system. Some systems theorists, predominately Checkland (1981), argue that to view the Universe as composed of discrete identifiable systems is inappropriate. Checkland warns that, often, the tightly technical application of the word 'system' as a description of a constituent part of a whole is often misapplied as a description of the whole itself. The best that can be achieved, if Checkland's contention is accepted, is that a phenomenon, such as a state run health service, cannot be viewed as a system, but *as if it was* a system. Many people find this a difficult concept to grasp, not least because it is very difficult within the literature to identify whether, when an author uses the word 'system', he/she means a 'true' micro view system or a gestalt, viewed-as-if-it-were-a-system, system. To get round this problem, Checkland and Scholes (1990) suggest using the word, 'Holon', coined by Koestler in 1967. For ease, in this text, we shall also use the words 'holon' or 'holonic' when referring to gestalt systems. Checkland and Scholes argue that the label 'holon' can be applied to a whole that exhibits, what they term as, 'emergent properties'. These emergent properties include a layered structure, process of communication and process of control that allow the holon to survive in a changing environment.

So, to illustrate, when we refer to the 'health service system', we are including all the participants of the overall service who work synergistically in different disciplines to maintain that service. We do not split the service down to each of its contributory systems; we view it as a whole. When we schedule a patient for theatre, we do not say that they have appointments to see the ward staff, the porters, the anaethestist, the operating department technicians and so on; when we say 'that the patient is going to theatre' we understand that all these other people are involved—we view it as if it were a system. We understand that it is a gestalt, greater than the sum of its parts and we assume that others also understand this. There are communication processes inherent within the health service system and processes that control the development and direction of health care in order to survive changes in the overall environment in which it functions. It is a holon.

Likewise, Lorenz's work modelling weather systems is viewed in a holonic rather than a micro fashion. To model a weather system in its strictest technical term would mean investigating all processes (systems) that contribute to the overall phenomenon known as 'weather'. Lorenz's work focussed upon changing one or two of twelve variables programmed into the computer to see how weather patterns would be affected. This is a good example of viewing an overall system holonically

Reed and Harvey (1996, cited by Byrne) refer to 'nested systems' that can be seen as characteristic of the whole system. This resonates to some extent to the notion of a layered structure as outlined by Checkland and Scholes (1990). Reed and Harvey's stance implies that the characteristics of the whole can be extrapolated from the characteristics of the contributory system. These nested systems are seen as microcosms of the whole. So if you see our patient being examined by an anaesthetist and then signing a consent form, you understand that this patient has surgery scheduled and the whole process is clear to you. The anaesthetic examination can be seen as a nested system. The joint approaches of Checkland, and Reed and Harvey are important because, if we are going to use Chaos Theory in a practical way, they help us to make sense of the whole concept of 'systems'. To clarify this, let us take the respiratory 'system' as an example (see *Table 1.1*)

Table 1.1 begins with a clear statement that defines our holon. It then goes on to outline the viewpoints of Checkland, and Reed and Harvey. Where they agree is that there are components or characteristics of the holon, which allow us to understand how it functions. When we start to use Chaos Theory to evaluate a specific nursing service, we will be using a combination of the approaches of Checkland, and Reed and Harvey. We will begin accepting that, at one level, the best we can hope to achieve is if we view our service as if it were a system, as a holon. We will accept that it has properties as a whole that might not be apparent if we examined every tiny contributory component of the service. Taking a macro view of the service means that we would not have to subject the service in question to extensive and exhaustive analysis. However, this holonic thinking is not incompatible with Reed and Harvey's approach. By identifying the 'nested systems' that make up the majority of the service and targeting them for evaluation, we will be able to draw conclusions about the future of the service, as we did with the anaesthetic examination earlier. The 'nested systems' that require our focus should, by virtue of their significant role, be immediately identifiable as crucial.

Table 1.1: Understanding holons

Holon definition	
The term 'respiratory system' refers to the group of specialised organs whose specific function is to provide for the transfer of oxygen from the air to the blood and of waste carbon dioxide from the blood to the air. The organs of the system include the nose, the pharynx, the larynx, the trachea, the bronchi and the lungs (Weller, 1989)	
• Checkland would argue that you could not use the term 'system' in relation to the physiological structures outlined above. What we define as the 'respiratory system' is in fact a gestalt of many systems, which all contribute to respiration. However, when we refer to the structures and processes involved in respiration, we do not have to mention every system involved. Instead, we view respiration *as if it were* a system • Checkland speaks of 'emergent properties', which allow an entity to exhibit properties as a single whole.	• Reed & Harvey would concur with Checkland's view that respiration depends upon many systems to be effective. However, they would also suggest that these systems are inter-related and that understanding one or two of them will allow us to speculate about the entire process. So, if we understand how the alveoli work, we can extrapolate that knowledge into an complete understanding of the entire 'system' • Byrne (1997) has suggested that only a small number of variables (i.e. one or two nested systems) will control the end state of a system

Holon parameters

A fundamental characteristic of a dynamical system is change. In mathematics, equations that express rates of change are referred to as differential equations. The rate of change is ascertained by the difference in values either in temporal or spatial (time or space) parameters (Stewart, 1997). In *Table 1.1*, we have identified our respiratory holon, thus:

> *'The term respiratory system refers to the group of specialised organs whose specific function is to provide for the transfer of oxygen from the air to the blood and of waste carbon dioxide from the blood to the air. The organs of the system include the nose, the pharynx, the larynx, the trachea, the bronchi and the lungs '*
>
> (Weller, 1989)

By identifying the specialised components within our gestalt system, we have defined our holon within spatial parameters. We could look into the chest cavity and recognise the components of our holon. So now, when we refer to our respiratory holon, we know exactly what we are talking about. Our holon does not include consideration of skeletal structure or skin integrity, even though they contribute to respiration; the parameters are clearly defined.

Any change in the parameters of a holon is a fundamental element of Chaos Theory. It is encapsulated by the phrase 'sensitive dependence upon initial conditions'. It suggests that, if you pick two different starting points in a system, no matter how closely they can be plotted mathematically; they will give rise to two different paths that will, eventually, diverge.

The importance of defining holon parameters cannot be over-emphasised. As a way to ground these abstract concepts in clinical practice, let us imagine that we want to compare two specialist nursing services in identical areas. The specialist services are our holons. Let us assume that both our holons deal with the same number of patients who have had the same kinds of treatment; both holons have an equivalent number of beds and resources. The holon parameter we are interested in is temporal; we want to find out how much patient contact each of the services provide over a specified time period. If we wanted to plot this on a graph (see *Figure 1.1*), as long as our two services remain identical, the trajectories would also be identical. However, if we change the holon parameter of one service, by removing one member of staff, for example, the trajectories would diverge over time until they were completely different.

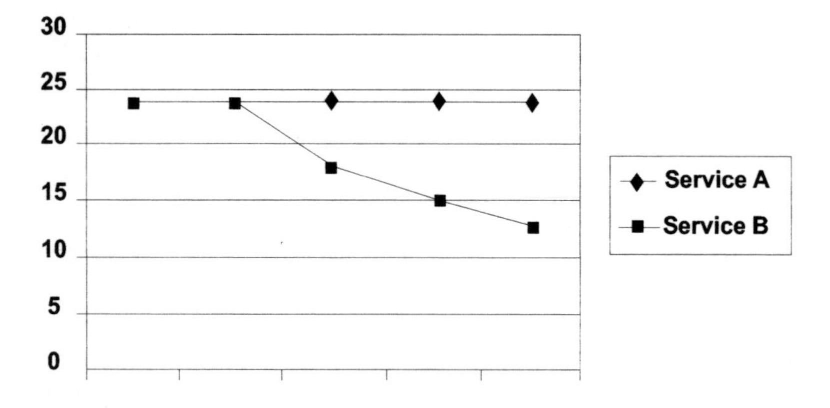

Figure 1.1: Changing holon parameters

Of course, *Figure 1.1* only illustrates one aspect of the overall holon, that of patient contact. It cannot give us any real insight into issues, such as quality of care, stress levels or staff/patient communication strategies, although we could infer this information experientially. *Figure 1.1* is indicative or predictive rather than explanatory. However, it does show very clearly the effects upon service provision if the holon parameter is changed very slightly. When we begin to apply the mathematics of Chaos to practice, manipulation of holon parameters will be fundamental; they will show us a future.

Equilibrium states

We have established that a dynamical system moves through space and time and changes as it does so by losing energy. When a dynamical system has reached a state of minimum potential energy, it is said to be in an equilibrium state. A dynamical system can have two sorts of equilibrium states, periodic and aperiodic. Thus, it can be seen that a dynamical system will eventually settle into one state or another; it will be stable (periodic) or unstable (aperiodic). It may appear strange to describe an equilibrium state as unstable, since 'equilibrium' typically describes a balanced state that remains unchanged as time advances. To illustrate this, let us take an example from Lorenz (1993).

Take a very well sharpened pencil and try to balance it vertically on the sharpened end. You can't do it, can you? The pencil may balance for a microsecond, but it soon becomes unstable and falls to the desk. This is because humans cannot judge sufficiently accurately how vertical the pencil is. Your eye and your hand are not discriminating enough to differentiate between a pencil that is truly vertical and one that is very, very, slightly out of alignment. This slight disturbance of alignment is enough to move the pencil out of its brief interval of equilibrium (does this sound like sensitive dependence upon initial conditions? It should!) The equilibrium of the pencil is unstable. Once the pencil has fallen onto the desk and come to rest horizontally, its equilibrium is stable; it will not move unless some outside force, like you knocking it onto the floor, changes its resting state. Once a dynamical system has reached equilibrium, whether periodic or aperiodic, it manifests itself as an attractor.

Specific state attractors

Stewart (1997), rather unhelpfully, suggests that an attractor is whatever a system settles down to be. However, it is true that the end state of a system will dictate what sort of attractor the system will become. An attractor can also be defined as a mathematical mapping concept that allows the geometry of a chaotic system to be represented graphically. This means that, when you have calculated the maths of a Chaos system, you can easily represent them as a graph; this will be extremely useful to us later on. Sardar and Abrams (1999) suggest that the cultural equivalent of attractors would be tribes, states and things that give us identity, like religion or class. Therefore, it can be seen that the attractor state of a system is something that helps to make that system unique and to give that system identity.

The simplest attractor that a system can settle into is a fixed-point attractor. In this case, the system will stop at a fixed point. For example, if you drop a stone from your out stretched hand, gravity will bring the stone to a stop when it hits the ground. The stone will not continue to drop through the floor; it has reached equilibrium. The dynamical system that was the stone moving through a gravitational potential becomes a fixed-point attractor.

The second example of a periodic equilibrium state is that of the limit cycle attractor or the closed loop attractor. The clearest example of a closed loop attractor is that demonstrated by the Volterra two-species differential model of predator/prey populations (Kosko, 1993; Cohen and Stewart, 1994). This is another thought experiment, like Schrödinger's dead/alive cat. However, its theoretical and conceptual base is not quite as difficult as Schrödinger's cat and it is quite easy to do. Think of an isolated island (no outside influences) that is home to two different species of animals, one of which, the predator population, lives by eating the other, the prey population. Basically, Volterra postulated that, as prey population rises so will predator population, until the balance tips and the prey population begin to fall as a result of over-successful predator breeding. As prey population falls, so will predator breeding rates until the original scenario is replicated and the cycle can begin again. As long as all the other parameters remain unchanged, the predator/prey population will move around in this cycle forever.

Both fixed-point attractors and limit cycle attractors are manifestations of periodic stable systems. When their geometry is mapped, their contours tend towards classical shapes and you can predict what is going to happen. The converse is true of aperiodic equilibrium. Aperiodic equilibrium in a system means that the system will appear to wander through its state at random. The slightest change in the system state will cause the state to diverge. This type of attractor is referred to as a strange or Chaos attractor. There is no classical shape to a Chaos attractor; it is more or less impossible to predict where it is going and its randomness when plotted gives rise to fractal geometry like those below.

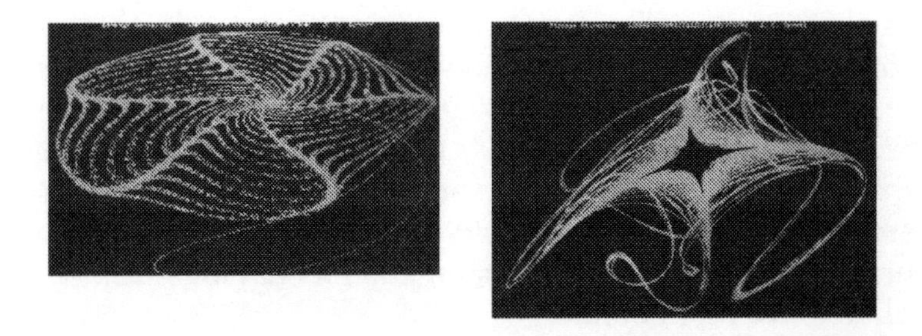

Sprotts Fractal Gallery—Reproduced with permission (Sprott, 1995)

Don't panic. Although we will be identifying Chaos from graphs, you won't have to produce anything like these fractals.

Although the chaotic system may diverge from a classical pathway, two important things must be borne in mind. Firstly, despite the system appearing 'random', its behaviour is deterministic and can be predicted via mathematical formulae. Secondly, the system will never diverge to the extent that it leaves its attractor state. Therefore, a fixed-point attractor will never develop into a limit cycle attractor and a limit cycle attractor will not suddenly turn into a strange attractor. This is important to us because it means that once we have identified the attractor state of a specific nursing holon, providing the holon parameters remain unchanged, the state will not change at some future point.

Nonlinearity

We have one last concept that is closely allied to Chaos Theory to grapple with, that of non-linearity. Chaos is nonlinear in nature, although nonlinearity does not necessarily imply Chaos. Lorenz (1993) highlights that the term 'nonlinearity' has become synonymous with Chaos, but states that this is an oversimplification of the facts, although he accepts that Chaos demands non-linearity. A linear equation means that any change in any variable, at any point in time, will produce the same effect size further down the continuum. Data from a linear equation, when plotted graphically, will produce a straight line—hence the name. Linear relations allow us to predict what will happen within a system—they form a straight line and we can easily identify where that line is going. Often, linearity is easy to detect in closed automatic systems that are programmed to behave in certain ways. Expecting linear behaviour to be seen in any holon that is concerned with people is probably a little unrealistic—people are unpredictable and inconsistent; their behaviour cannot be plotted in a straight line.

Non-linearity, however, is a mathematical representation of the, now familiar, chaotic concept, 'sensitive dependence upon initial conditions'. It introduces uncertainty into our calculations. In other words, it introduces Chaos and Chaos is not linear. If we return to the butterfly effect, for example, Lorenz suggests that if chaotic systems were linear, the flapping wings of the butterfly may produce a 10-degree rise in temperature. If we then postulate a disturbance of air current 100 times greater than that of a butterfly—that produced by a seagull, for example—we would expect a temperature rise of 1000 degrees, which is clearly nonsense.

To illustrate these concepts from a clinical perspective, consider the following examples. If you buy a box of dressings for your clinical area that cost £100, someone who buys two boxes of the same dressings will pay £200—that is linearity. Knowing that, means it is really easy to work out how much 5 or even 500 boxes of dressings will cost. However, suppose you want to work out how much food to order for your nursing home. You might take a linear approach and order as many meals as there are occupied rooms, but this does not take into account the different appetites of the residents. Appetite can depend upon physical health, gender, activity levels, etc., and these are not always predictable. If a resident who weighs 40 kilos eats two sausages for breakfast, it doesn't mean that another resident weighing 80 kilos will eat four because appetite is non-linear. Introducing these non-linear concepts changes the deterministic rules within the system. When you start to examine the service you provide, you should quickly be able to detect the non-linear nature of your work because real life is 'messy' and, to varying degrees, uncontrolled.

Summary

This chapter has introduced the elements of Chaos Theory that we need to understand if we are going to apply the principles of Chaos Theory in the management of specialist nursing services. *Table 1.2* reiterates the most important concepts.

Table 1.2 – Important concepts

- Chaos can be found in dynamical systems that change over time
- A small change in a dynamical system will give rise to significant effects later. This is referred to as sensitive dependence upon initial conditions
- Chaos is non-linear
- To distinguish 'gestalt' systems from 'real' systems, we refer to gestalt systems as 'holons'
- Holons have layered structures (Checkland and Scholes, 1990) and nested systems (Reed and Harvey, 1996); these allow us to make conjectures about the whole
- Manipulation of holon parameters will allow us to predict likely trajectories, which, in turn, will show us the end state of the holon
- The end state of a dynamical system or a holon will determine what its specific attractor state will be
- A holon has the potential to become a fixed-point attractor, a limit-cycle attractor or a Chaos attractor
- Once a holon has settled into an attractor state, and *providing the holon parameters remain unchanged*, it will never leave it
- The attractor end state can be plotted mathematically
- Once we have plotted the attractor end state, we can make predictions about service demand/delivery
- These predictions only hold true if holon parameters remain unchanged.

These are important concepts with which to get to grips. If you aren't clear about what they mean, you will find it difficult to put the Chaos Theory into the context of your practice. The sorts of questions you might want to ask yourself are:

1. Can I recognise a dynamical system (one that changes over time)?

2. Am I clear about what can cause a change in the direction of that system?

3. Do I understand the difference between the various end-states that a dynamical system can achieve?

4. Can I recognise these states if I look for them?

5. Am I clear on what a 'holon' is?

6. Can I recognise nested systems within the service I offer?

Now that the important concepts of Chaos have been introduced, the issue is how to utilise them in a way that can help manage nursing practice. Chaos is not a new concept in health care, as we will see. Unfortunately, nursing appears to have had some difficulty in applying the tenets of Chaos in an informed or useful manner. The next chapter will provide an overview of the uses that Chaos Theory has been put to in the health care professions.

2

Reviewing Chaos[1]

The science of Chaos has been utilised in many disciplines that contribute to health care. It is not overly fanciful to suggest that Chaos Theory is, itself, a dynamical system and that the application of the theory has changed focus over time. The earlier days of Chaos thinking were characterised by an almost exclusively physiological focus. However, as the discipline evolved, Chaos thinking was applied to a variety of disciplines as diverse as epidemiology and economics. By the twenty-first century, American nurse theorists were applying its principles to the organisation and evaluation of care delivery. In general, the application of Chaos Theory to nursing has fallen into one of three categories: 'pseudo-science', research or statistical modelling. It can be argued that these domains demonstrate the two extremes of Chaos philosophy, the mystical and the empirical. The aim of this chapter is to explore the different applications of Chaos Theory that various health disciplines have attempted in order to demonstrate to the reader that, although Chaos has often been imperfectly understood and inadequately applied, when used by informed individuals, it has a significant role to play in the management of practice.

Chaos Theory has been applied in a practical manner by various disciplines, such as the neurosciences, cardiology and social sciences. However, nursing has not been particularly successful in its attempts to use Chaos in such a pragmatic way. Chaos Theory, as an investigative, explanatory and

1 Some of the content of this chapter appeared in the following article, Haigh C (2002) Using Chaos Theory: the implications for nursing. *Journal of Advanced Nursing* **37**(5): 462–69, and is reproduced here by kind permission of Blackwell Publishing

predictive tool, has not been well served by the (predominantly American) nurse theorists who have espoused its cause.

Chaos as metaphor and pseudoscience

Writing in 1997, Vincenzi noted that the application of Chaos thinking was changing. There was less emphasis on finding Chaos in various phenomena, and more focus on using Chaos techniques to understand the world of nursing. In this approach, there is no real need to identify the chaotic elements of nursing (if, indeed they exist); rather, the presence of Chaos is taken as a given 'fact' without troubling to provide data to support such an assertion. Hayles (2000) refers to this approach as a 'metaphorical approach', and suggests that chaotic concepts—such as sensitive dependence upon initial conditions—are presented as truths, but should really be viewed as metaphors that illustrate the phenomenology of nursing. This suggests that essential chaotic truths may illustrate nursing practice, but do not necessarily form a fundamental part of it. She further suggests that adherents to this metaphorical approach tend to 'blur the lines between metaphor and analysis' (p3). Hayles acknowledges that this metaphorical approach will free researchers from seeking Chaos, and permit them to predict system dynamics. However, she warns this may be a dangerous approach, since there is no empirical evidence that the assumption that Chaos dynamics exist within nursing is a reasonable one. Data gathered from a flawed perspective will be neither meaningful nor useful. This is a good point but, in the context of this book, its validity can be questioned. The focus of this volume is the application of chaotic principles to analyse service development; however, we make no assumptions regarding the existence of Chaos; indeed, the quantitative evaluation we will carry out is focussed upon seeking and recognising Chaos, if it exists.

Hayles makes the point that nurse theorists who wish to integrate positivist techniques with 'softer' elements of nursing research use this metaphorical approach. The result is often a blend of the 'scientific' and the 'mystical', which serves to weaken both the application of Chaos Theory and the status of nursing research. This point is well illustrated by the work of Ray (1994).

Ray sees nursing as a complex dynamical system and argues that four fundamental dynamic processes can be identified, contributing to the concept of nursing enquiry (which, she explains, means the history-taking element of nursing care). She further postulates the existence of 'caring attractors', which one can only assume (since she does not clarify her thinking) are somehow the nursing equivalent of chaotic attractors, as described earlier. Caring is an extremely abstract concept and one that has never been satisfactorily defined. Ray does not elaborate how the concept of 'caring' can be mapped mathematically, which significantly weakens the 'caring attractor' argument since, as we will see in subsequent chapters, the whole point of identifying an end state attractor is to plot graphically how that attractor looks and where it is likely to go. In addition, she suggests that chaotic systems are unpredictable, which is patently a false premise, Lorenz (1993), Gleick (1987) and countless other authors have clearly stated that Chaos is the term that is given to the underlying predictive order in seemingly random systems. However, she not only suggests that accepting a chaotic dimension in nursing is necessary to decision-making in nurse-patient relationships, but also highlights that this element of Chaos allows, 'Interconnected pattern-recognising goals sustained by cosmic powers of space, time and life' (p30). Such metaphysical approaches do not always sit comfortably within the mindset of UK health care. Ray attempts to adopt a positivist stance by continually referring to scientific theory; however, she weakens this stance by spurious claims for a 'new scientific theory', which sees life processes as mental processes, suggesting interconnectedness between mind and matter. Overall, the work leaves the reader with the impression that Chaos Theory is being used as a crutch to give some 'scientific' credibility to a metaphysical analysis of the discipline of nursing. There is no indication that the author has any insight into the requisite elements of Chaos thinking. The focus of the work is so deeply impressionistic that the incorporation of any positivist thinking is redundant.

Similiar concepts can be found in the work of Sabelli *et al* (1994), who regard the 'oneness' of the individual with the universe as a point attractor. They take the stance that a biopsychosocial paradigm is an extension of the biophysiological approach typical of systems theory, and argue that the manifestation of system equilibrium is health, while illness is manifested as a chaotic attractor. There is no evidence that even the most fundamental understanding of the components of chaotic thinking are understood or applied here. In an illustrative case study, Sabelli *et al* introduce the concept of the 'butterfly effect' when examining the psychological effects of stress on cardiac activity. However, they do not demonstrate how these psychological effects can be mathematically plotted to identify end state attractors. Once again, the spiritual and mystical view of the nature of humans and their place in the universe is at the heart of the work. This may reflect the condition of American nursing theory, which has often sought the more esoteric side of nursing and attempted to incorporate it into mainstream nursing care. To the uninformed, Chaos Theory, with its method of demonstrating underlying order and predictability in superficially disordered systems, might appeal as a way of seeking some spiritual understanding and overall plan for the nature of life and the universe.

Such paradigms resonate with the work of theorists like Rogers who, in defining the environment as a four dimensional energy field composed of waves that evolve towards increasing complexity and diversity, incorporated an incomprehensible amalgamation of Chaos Theory and the second law of thermodynamics into her work (Rogers, 1980; Fawcett, 1984). Although influential in American nursing circles, Rogers has been justly criticised in the UK for using science in inappropriate ways and undermining the credibility of nursing knowledge (Johnson, 1999b). As a vociferous critic of what he terms 'pseudoscience', Johnson warns of the dangers of the application of scientific principles in inappropriate and uninformed ways. He also counsels approaching with caution works that liberally sprinkle the names of famous theorists throughout the text with no compelling evidence that the author has even the most basic familiarity with the works quoted (Johnson, 1999a). Johnson invokes the notion of a 'five minute test', which suggests that any writer should be able to speak coherently for five minutes about the author of any work they have quoted and their influence in the wider context. It is difficult to imagine that some of the more esoteric authors cited within this chapter would manage to pass the Johnson five minute test. A further concern is noted by Hayles:

> *'One of the reasons that Chaos Theory lends itself to application in so many different disciplines is that the analysis of chaotic data does not require a detailed understanding of the mechanisms involved"*
>
> (Hayles, 2000: 1)

While acknowledging the point that human knowledge increases because individuals strive to understand the unknown, it is difficult to imagine a quantum physicist, for example, embarking on any project without a detailed understanding of the fundamental mechanisms involved. In the interests of balance, it must be noted that Niels Bohr, the man who is widely acknowledged to have had the greatest influence on quantum physics is quoted as saying, *'Anyone who claims quantum theory is clear, doesn't really understand it'* (Strathern, 1998). Nonetheless, it can be argued that some basic knowledge of the theory that drives Chaos is essential to the informed application of chaotic elements of analysis.

This contention, that lack of understanding does not necessarily stand in the way of a theorist's application of Chaos to health, is also illustrated by the work of Pediani (1996) and Walsh (2000). Pediani bravely attempts to identify elements of Chaos within the concrete nursing issue of poor analgesia administration by nurses. Although his premise is never clearly articulated, his hypothesis can be inferred from the early part of the paper, which uses Chaos to gain insight into the basic features of the dynamics of complex areas. Lodged in the metaphorical approach outlined by Hayles, Pediani assumes that chaotic elements are present in the basic nursing task of ensuring adequate analgesia administration to the patient without actually attempting to demonstrate them. He implies that an understanding of these elements will provide insight into nurses' concerns regarding iatrogenic addiction. However, before embarking upon any consideration of the chaotic elements of the issue in question, Pediani arbitrarily abandons his task in favour of a consideration of meme theory. Meme theory concerns itself more with the evolution and transmission of concepts and ideas than systems analysis. While it can be argued to contribute in some cases to chaotic systems, meme theory is not in itself a part of Chaos Theory (Dawkins, 1989).

Walsh (2000) aligns himself with the stance of the American theorists previously considered, in that he argues strongly that Chaos Theory can contribute to a holistic consideration of individual patients. He rejects the notion of conceptual role boundaries, suggesting that they are artificial, without bothering to acknowledge that most practitioners will, nevertheless, have to function within them. Although he acknowledges that these boundaries may be blurred in some instances, he does not develop this thinking towards the identification of attractor states. Furthermore, Walsh introduces philosophical concepts, such as the nature of the life/death divide into his argument, which do not sit comfortably with the empiricist nature of Chaos thinking.

Chaos as a research tool

The nursing profession seems to be having difficulty in applying chaotic thinking to reality-focussed clinical practice. Therefore, the contention of Mark (1994) that the major role of Chaos thinking is not necessarily in health care delivery, but in nursing systems research, does not appear inappropriate. She suggests that, while we can incorporate the notion that nursing is a dynamical and chaotic system into our thinking, it is as a tool for examining system parameters that it excels. She suggests that Chaos thinking can help a researcher to identify the attractor state of a particular aspect of a nursing care system, and predict service stresses as the system moves towards unstable equilibrium. She acknowledges that this approach to

Chaos theory use offers enormous methodological challenges to researchers, at the same time arguing that the data obtained would be of enhanced richness and would reflect the non-static status of care organisation and provision in a valid and reliable manner.

The weakness in Mark's argument is that at no time does she acknowledge that the data sets, which facilitate the identification of chaotic systems are not yet available to nurse researchers. The development of such data sets would require the careful use of a multiplicity of research methods and systematic validation, before the wholesale categorisation of dynamical systems of nursing could even be considered. The generation of these data sets would be wholly reliant on the identification, formatting and manipulation of theoretical, conceptual and contextually dependent boundaries—all of which would be required before the notion that nursing is composed of dynamical and chaotic systems could be seen as a viable assumption. Strathern (1998) proposes that 'plurality is the belief in the coexistence of incompatible views that has gradually permeated all aspects of twentieth century culture', which suggests that multi-method approaches to data set production should be easy and acceptable. However, this is not the case. The difficulty in implementing this approach to data set production reflects the negative attitude of nursing research, in particular, to using research methodology in a less than 'pure' way. Johnson *et al* (2001) make a persuasive case for plurality in qualitative health research. By plurality, they mean that researchers should use a multiplicity of methodologies and intellectual approaches in order to gain a richer picture of the subject they are investigating. It is unfortunate that they do not include a similar consideration of quantitative approaches in their paper, since the argument for pluralistic approaches, such as those which they outline, can also be made in the context of identification of chaotic data sets. This is because the collection of such data may require the implementation of qualitative methods, such as focus groups or interviews, to clarify the nature of a specific nursing system, before individual elements of that system can be examined using quantitative methods of evaluation. The data obtained for this quantitative approach will, in turn, contribute to the identification of chaotic attractors within the system. Haigh (2000) used this pluralistic approach when applying Chaos Theory to the estimation of potential longevity of a specific specialist service. Qualitative methods, such as focus groups and non-participant observation, were used to identify key nursing systems which, in turn, were subjected to mathematical analysis, allowing chaotic attractor states to be plotted.

Elaborating on this concept, Byrne (1997), writing from a social science perspective, cited Trisollio's contention that complexity is a representation of 'the sharing of ideas, methods and experience across a number of fields' (p1). Byrne argues that the plurality of Chaos serves to highlight the limits of conventional science in describing the world. He attacks the reification of quantitative methodologies and suggests that the quantitative can be viewed as inherently qualitative, and that, within the social sciences, the generation of anything other than very general non-contextual laws is impossible. Copnell (1998) suggests that, even when apparently divergent methodologies are used, the underlying belief systems of the researcher remain the same and further supports this stance. Application of non-linear thinking in nursing will require the plurality of approaches espoused by Johnson *et al* (2001), coupled with the multiparadigm approach highlighted by Copnell (1998). However, the contention that transitory laws can be drawn from data resonates with attitudes within what are traditionally viewed as the 'hard' sciences, such as physics, where the view is held that the notion that all scientific laws are immutable is nonsense. Scientific law is constantly under review, and constantly altering in the light of new knowledge.

This supports the work of Coppa (1993), who submits that the role of Chaos Theory in nursing should be viewed within the framework of Kuhn's views about paradigms, normal science and revolutions. Coppa argues that Chaos Theory can be used to build a new paradigm of nursing science that would facilitate the integration of practice and research. Contexualising this within the work of Kuhn (1970), Coppa maintains that the boundaries of the existing nursing paradigm are relaxing, the nature of nursing research problems are changing, and the science of nursing is in a pre-revolutionary or pre-paradigm stage. The thrust of her argument is that using the concepts of non-linear dynamics will allow nurses to move outside the traditional philosophies of their science, and that nurse researchers will be freed from reductionist viewpoints and constraints of order and predictability.

In this, Coppa can be seen to be misleading. She adheres to the notion that reductionism is in some way 'bad', and that order and predictability have no role in Chaos thinking. Byrne (1997) notes that one way of analysing a system from a chaotic perspective is the construction of time-ordered classification within data sets. This addresses the issue of the manipulation of contextually-dependent boundaries, since all systems in the physical world operate in a time-dependent capacity. We introduced Reed and Harvey (1996) in *Chapter 1* with their 'nested systems' that can be seen as characteristic of the whole system. The major assumption in Reed and Harvey's work is that the characteristics of the whole can be extrapolated from the characteristics of the contributory system. The argument is that nested systems are seen as microcosms of the whole, which can be evaluated as chaotic attractors and will facilitate statistical modelling of care. This can be argued to be an appropriate use of Chaos Theory, and one that may be attractive to nurse researchers. However, such statistical modelling is reductionist in nature and diametrically opposed to the stance of Coppa (1993), namely that Chaos Theory will free nurses from reductionism.

Statistical modelling using Chaos Theory

Gleick, (1987) notes that it was the 1980s that heralded the development of Chaos Theory in the field of physiology. Doctors and physiologists were viewing the body as a myriad of dynamical systems and identifying elements of Chaos within all of them. Elements of Chaos have been identified in respiratory rate (Petrillo and Glass, 1984) as well as within neural mechanisms (Glass and Young, 1979).

Non-linear dynamics have been closely associated with the field of physiology for some time. Mathematical models of the heartbeat were first formulated in the 1920s. Van der Pol's model replicated his earlier work calculating the oscillation of an electronic valve, which Cohen and Stewart (1994) suggest as an example of a limit cycle. However, the heart is not the only organ in the body that is controlled by electrical impulses. Techniques for monitoring the electrical impulses of the brain via electroencephalogram (EEG) recordings have been established since 1929 (Cohen and Stewart, 1994).

The effectiveness of such modelling in the neurosciences has been shown by Martinerie *et al* (1998). Martinerie and colleagues suggested that deterministic non-linear processes are involved in the pre-seizure cerebral neural reorganisation that is a precursor of an epileptic attack. The implication of this is that, if the processes involved in initiating an epileptic seizure can be plotted using non-linear procedures, pre-seizure prediction would be possible. The application of non-linear analysis was facilitated by the use of EEG, which allowed the dynamical system of a patient's neuro-electrical activity to be studied. EEG recordings are time-series studies, allowing the mapping of attractor topology and the potential identification of Chaos within the system. Martineire and his colleagues analysed 19 seizures from 11 patients and identified the changes in cellular excitability that reduced the threshold for seizure onset. This pre-ictal (pre-seizure) process was facilitated by the recruitment of distant neurons in epileptic prone tissue. When the numerical data was plotted, the trajectory of inter-seizure and seizure neurological transitions showed congruence in their geometry. In practice, this suggests a 'route' toward neuronal recruitment and patient seizure, which can be mapped and predicted. The 'sensitive dependence to initial conditions' of the neuronal dynamical system is represented by the proposal that other patient specific processes also affect this route. In this case, the recruitment of pre-ictal neurons can be viewed as a nested system, in that their characteristics allow for extrapolation of events to take place. The work of Martinerie *et al* shows that, given patient specific baseline data, seizure prediction 2 to 6 minutes beforehand was possible in 89% of their sample. Detection of ictal onset of this nature will allow for electrical stimulation, via subcutaneous implantation. This will divert the epileptogenic stimulus away from its seizure route. This is an excellent example of statistical modelling of chaotic end-states showing clinical application.

Rapp (1993) had already identified the confounding factor of 'noise' in Chaos measurement and had emphasised that, in neurosciences, the noise to signal ratio was very low and suggests that high resolution EEG systems be utilised when collecting neuronal data for the identification of Chaos. Although Rapp was concerned with extraneous neuronal activity that may affect the clarity of an EEG reading, the concept of 'noise' as external stimuli that should be recognised and discounted when seeking Chaos is one to which we will return when we start to seek chaotic elements within the organisation of nursing services. Rapp suggests that, while the traditional scientific approach to data collection is to sample as often as possible in order to give breadth and validity to conclusions, over-sampling can seriously bias analysis in favour of Chaos. This resonates with the work of May (1976) who suggested that too much data leads to less transparency in Chaos analysis. Rapp concludes that two of the weaknesses of Chaos Theory application, particularly in the biological sciences, are centred on sampling and system assumptions. Rapp notes that sampling criteria which try to avoid contamination by 'noise' are rarely developed. He also argues that, frequently, the assumption is made that the system under investigation is stationary; that is, it is assumed that the underlying processes producing the signal do not change during the period of measurement. This also can be seen to be true of nursing services. No service remains static, seasonal variation in admission patterns or changes in national health policy all contribute to turning any method of health care into a truly dynamical system. However, this confounding variable, or 'noise' need not be an obstacle to applying Chaos Theory, providing a) it is acknowledged to exist, and b) the end state of the system is recognised to be a indicative rather than a concrete and predictive definition of events.

Rapp's (1993) paper concludes by suggesting that the methods used to analyse may be more important than Chaos itself. As you will see when we start to look at how to seek

Chaos in nursing services, this is a valid point. The methodological approaches that we use to identify the 'noise' in our system and to validate the parameters we choose to set are as important as the end state attractor we identify. Rapp postulates that Chaos is only one manifestation of dynamical systems and the true value of Chaos is the mathematical techniques that have been developed to express it.

This would suggest that Chaos could be used to identify trends within a population or service. The rationale for this is considered by Hamilton *et al* (1994) when they suggest that very few of the scenarios that nurses encounter can be described as linear. If our care was linear, we would not see essentially equivalent patients responding in very different ways to the same health intervention; services would not have diverse levels of effectiveness in different geographical areas even when the target populations are demographically indistinguishable. Hamilton *et al* (1994) used Chaos Theory as a method of carrying out cluster analysis in the non-linear dynamics of births in adolescents. This application permitted the identification and manipulation of variables in order to carry out predictive modelling, and the identification of epidemiological markers that might contribute to the incidence of teenage pregnancy.

There are two key points in this work. Firstly, the application of time series analysis, seeing the holon of concern as a dynamical system set within temporal parameters and, secondly, Hamilton and colleagues also attempted to control for 'noise' by identifying their data as a low dimensional attractor state. This meant that the data was recognised as indicative of a process dependant upon only a few variables for the pattern of its change over time. When we start to seek Chaos within clinical practice, we also will be operating with low dimensional attractor states as our starting point.

This has been done before. Haigh (2000) used a nested system approach by evaluating the patient contact element of an acute pain service as a method of identifying the chaotic attractor status of the whole service. By treating patient contact as a time-series nested system within the overall system, she was able to model chaotic end points via parameter manipulation that provided insights into the evolution of the entire service. Haigh's work forms the basis of the illustrative case study found in *Chapter 4*. Griffiths and Byrne (1998) have suggested that only a small number of variables will control the end state of a system; Hamilton *et al* (1994), Martinerie *et al* (1998) and Haigh (2000) have all shown this to be the case.

Griffiths and Byrne (1998) have likewise postulated that statistical modelling of nested systems can contribute to mapping the perspective of patient outcomes in general practice. Ireson (1998) has suggested that this type of modelling has a function within the organisation and management of nursing care. The work of Hamilton *et al* (1994), applying Chaos Theory to cluster analysis, supports the contention that the statistical modelling of health care systems, rather than in research methodology, may be the ideal niche for Chaos Theory.

Summary

The purpose of this chapter has been to review Chaos Theory and to articulate the contribution that it might make to the discipline of nursing. When evaluating the application of Chaos Theory to specific aspects of care delivery, it is clear that the efforts made to use Chaos on specific clinical exemplars have been both ill-advised and ill-informed. This may well be because the conceptual fundamentals of nursing do not readily lend themselves to quantitative analysis, but could also reflect the pragmatic nature of nursing in general, and the problems that its practitioners might have in grappling with abstract concepts. For this to be clear in your own mind, you need to ask yourself the following questions:

1. Can I see how something that clearly has use in 'harder' sciences can be translated into a useful tool for evaluating health care holons?

2. When I review my own practice, can I identify 'noise'?

3. Am I clear what is meant by a low dimensional attractor and can I see how that applies to what I do?

4. Am I clear on how statistical modelling forms a framework for the identification of Chaos?

Many theorists, notably Coppa (1993) and Mark (1994), have suggested that Chaos Theory will contribute in a meaningful way to research disciplines rather than to practical nursing. This is a seductive notion, but it is clear that Chaos Theory is an ingredient of nursing research rather than a methodology in its own right. Copnell (1998) has, to a certain extent, pre-empted the case that Johnson *et al* (2001) presented for pluralistic approaches to research, in that she has disputed the use of Chaos Theory as a stand-alone research technique. She argues that, taken on its own, Chaos Theory cannot contribute to knowledge synthesis and, if it is to be an effective research tool, it must rely on a synergistic relationship with other research methodologies.

It is apparent from the work of Hamilton *et al* (1994), Byrne (1997), Martinerie *et al* (1998), Ireson (1998) and Haigh (2000) that the individual elements of Chaos can be identified within a wide range of health care systems—from social sciences to neurosciences, via epidemiology and nursing care management. These elements can be identified within parameters that are amenable to theoretical manipulation, and which contribute to statistical modelling. Such modelling allows for an appropriate use of quantitative methods, which may be blended with qualitative approaches in the setting of boundaries. This might allow for the identification of attractor states within services or systems that, in turn, could contribute to forward planning in a service or in a specific illness trajectory. In conclusion, it can be contended that for some years Chaos Theory has been viewed as an answer searching for a question. This search has not been aided by the uninformed application of the Theory by some writers. In the expanding area of systematic statistical modelling, Chaos Theory might, at last, have found the niche into which it fits. This, of course, is the stance taken by this book; certain aspects of Chaos Theory can be used in an informed way to indicate the end state of a particular health focussed service or aspect of care. Just how this can be achieved is outlined in the following chapter.

3

Population equations, parameter identification, data formatting and Chaos mapping

We have seen that the concepts of Chaos can be identified within dynamical systems and holons. By identifying our holon parameters and subjecting them to manipulation, we can predict the holon end point and detect the elements of Chaos within it. This chapter will explore how we can manipulate holon parameters by applying a simple population equation to predict service growth. We will explore the nature of the equation and apply it to a real world situation, using a specialist acute pain service as our data source, utilising a step-by-step approach that will clarify every element involved. In order to render all data comparable, the need and technique for data formatting will likewise be explored. The data used will be mapped graphically in order to identify the end state attractor of our model service.

The equation

The equation that we will use is applied in a population where generations do not overlap (May, 1976). This can be a seasonally breeding population such as temperate zone insects or, as in this study, a population of hospital patients. It requires a quantifiable 'seed' population, a clearly expressed rate of growth and a growth rate limit to be factored in. The equation itself is an extremely basic one that will be familiar to all population biologists:

$$x_{(next)} = rx(1 - x)$$

In this equation, x represents the population under examination and r represents rate of growth. Therefore, $x_{(next)}$, refers to next year's population. The element, rx, refers to the rate of population growth, which is multiplied by the seed population. The parameter, r, represents a rate of growth that can be set higher or lower. In this model, the final term is $(1-x)$. 1 represents the greatest possible population, so $(1-x)$ keeps the growth within boundaries since as x rises $1-x$ falls. So, to clarify, the equation:

$$x_{(next)} = rx(1-x)$$

means next year's population can be estimated by subtracting this year's population from 1 (to identify growth limits); multiplying the result by this year's population (because next year's population will be a function of x – this year's population) and multiplying by r (which is our rate of growth).

This is a modification of a Malthusian model. Thomas Malthus (1766–1834) theorised that a population increases faster than its means of support and that, unless it was checked by sexual restraint, it would only be limited by famine, pestilence, war, etc. Thus, this equation is a modification since, in a Malthusian model of unrestricted growth, the linear growth function rises forever upward. In our model, the term 1–x keeps the growth within bounds because of the x to 1–x relationship. In a straightforward Malthusian version of this equation, this term would be absent, thus:

$$x_{(next)} = rx$$

This would illustrate the traditional Malthusian scenario of growth, unrestrained by nutritional supply, environmental constraints or moral thinking (Gleick, 1987). It must be emphasised that this modification means that the equation under consideration introduces the concept of non-linearity into the work.

This nonlinear equation was considered by May (1976) as an example of a simple equation with complex dynamics. May carried out the first Chaos mapping exercise that was dependent upon manipulation of system parameters. By taking the population growth parameter and investigating the hundreds of potential values this parameter could exhibit, May was able to produce diagrams that represented the fate of the population under study. May noted that, when the growth parameter is low the population would become extinct, when the growth rate is higher the system begins to demonstrate instability (May, 1976; Gleick, 1987). This will be illustrated in more depth when we look at the effects of parameter manipulation in *Chapter 5*.

Gleick (1987) provides a clear recipe for the application of the population equation:

- Identify the starting population $-x$
- Identify the growth factor $-r$
- Subtract x from 1
- Multiply the result by x
- Multiply that result by r
- Repeat the process using the new population as a seed.

Thus, we have introduced an equation that will allow us to predict and plot the development of a specific population without overlap of generations. The equation has a factor that acts as a deterrent to unrestrained Malthusian growth. This will work well with tadpoles or minnows, but how can it be applied in a clinical setting? The answer lies within the careful definition of characteristic holon parameters and the identification and formatting of data.

Parameter identification

It has been emphasised that the non linear equation that we are planning to use, in order to evaluate the life expectancy of our service, has its roots in population prediction. However, the question we must ask at this point is, 'Can we really cover all aspects of our nursing care with one population equation?' And, of course the answer is 'no'. If we are going to use this equation, and gain some insight into our service, we have to do two things.

Firstly, we have to identify for ourselves what aspects of our care can be seen as 'nested systems'. Remember that Reed and Harvey (1996, cited by Byrne) suggested that nested systems are characteristic of the 'whole' system'. The implication here is that, once you understand the characteristic of the nested system, you can extrapolate the function or meaning of the whole. For example, if an alien spaceship landed outside your hospital and the aliens entered a general surgical ward, they might see various small tasks being undertaken. They might notice some people having wounds cared for. They would note that the people in bed seemed to be helped in daily tasks by the people in uniforms. They would notice that one person seemed to be in charge and was directing care, dealing with concerned relatives and liaising with other carers. We know that there is more to nursing than wound care, hygiene management and ward management but, as nested systems, they will allow our aliens to grasp the fundamental concept of nursing care. In order to acquire some meaningful data from using the equation, we must take a good hard look at the service we are planning to subject to analysis. The sort of questions that this preliminary examination asks mighty be:

1. What part of service provision takes up most of my time?

2. What part of service provision is most important?

3. When people think of the service I provide, what is the first task that comes into their minds?

4. What tasks do I do that most typify the nature of the service I provide?

Answering these questions will allow you to decide what parts of the service you are going to identify as nested systems. By deciding which parts of the service you provide best typify what you do, you are discarding extraneous variables; this is the 'noise' that we discussed in *Chapter 2*. This does not mean that 'noise' is unimportant, just that you have prioritised other aspects of what you do above it. For example, the bane of many lives is attending certain meetings; if you choose to categorise these meeting as 'noise' (not fundamental to the role you fulfil, but part of it), you can exclude them from your analysis, ensuring that only the elements of your function that you feel directly affect the patient experience are considered. You can use the data you get from this exercise as the basis of your seed population within the equation. Once you have done that, you have taken the first steps towards identifying the chaotic nature of your specific service.

The second step is to investigate the parameters of the nested systems you have identified. This will enable you to develop quantitative data that can then be plugged into our equation to facilitate prediction of service development. The types of question to ask when setting nested system parameters are:

1. What aspects of my service am I focussing on?

2. Have I identified those aspects of my function that are 'noise'?

3. How am I expressing this task; is it in hours spent or patients seen, or both?

4. How many hours do I spend on this task in relation to other aspects of my role?

5. How many hours do I spend preparing for this task in relation to other aspects of my role?

6. How many patients *on average* do I see in one hour/one week/one month (the time scale is up to you, but experience suggests that pts/hour is the easiest to calculate)?

As with the earlier examples of questions to ask, you should not use this as some sort of checklist for service exploration. These question are just examples of the focus of inquiry to get you started.

Once you have carried out these tasks, you will have some preliminary data that will allow you to start using the equation. You will have identified your 'system' or holon by identifying the nested systems within it. You will also have generated some seed population data that can be plugged into the equation as a starting point. All that remains is to identify changes in the nested systems over time to give an indication of service growth rate. This can be done retrospectively, by tracking back through your own records or those held centrally by your service, or prospectively, by collecting specific data over a span of time. However, if you opt for the prospective approach, it may be up to a year before you can start to calculate Chaos. In light of this, the retrospective approach might be more useful to you. You could also find it helpful to invite a person external to the service to help you with some of this. For example, if you want to check how long you spend talking to individual patients, it will be more accurate (and less intrusive) if a third party is clock watching and recording timings for you.

Data formatting with a single nested system

Thus, we have established that to use the equation, which was originally designed to predict the effects of population growth in biological environments, we must be very specific about the identification of our holon parameters. To illustrate this, we will take the example of an acute pain service as our holon (see *Table 3.1*).

Table 3.1: Pain service holon parameters

The Holon: An Acute Pain Service (APS) in a medium sized (800 beds) general hospital that services both urban and rural areas. The APS has been in place for eight years. While primarily servicing the surgical unit, the Acute Pain Nurse (APN) attached to the service also visits other areas to advise on pain management techniques. Additionally, the workload also carries a large teaching remit, participation in audit and research and day to day service management.
Nested system identified: All patients who had used patient controlled analgesia (PCA) post operatively
Identifying the seed population and growth parameters : We need data for two years

- ***Year 1***: 464 patients (seed population)
- ***Year 2***: 619 patients
- ***Growth factor***: 25%

Firstly, we must identify a specific nested system that is fundamental to the efficient running of the holon. The major stakeholders in the service, in the case of our example—the acute pain nurse—should identify the system parameters. For this example, the APN in our example has suggested that the number of patients seen over the year best represents the service. So, we will take our designated population as all patients who have used patient controlled analgesia (PCA) after surgery and agree that all other aspects of the service be designated as 'noise'. We also need to be able to calculate how the population has grown over a cycle of time. May (1976) noted how the non-linear equation that is the focus of this work was applicable to any population where generations did not overlap. While it is acknowledged that this is difficult to guarantee in a hospital patient population, it can contribute to decision-making when deciding what time period to use. Providing one can identify a population growth factor, there is, in theory, no reason why population data cannot be compared on a month by month basis rather than a year by year one. However, in that scenario, it is unlikely that patient 'generations' will not overlap and this will compromise the quantitative data. It is acknowledged that, even in a year-on-year comparison, there may be some generation overlap; however, it is contended that this will be minimal and less of a confounding factor than it would be in a month-on-month comparison Therefore, we need to have data on the PCA patient population for two years. *Year 1* gives us our seed population of 464 patients. By *Year 2*, this has risen to 619 patients, an increase of 25%. So, we have the numerical data for the first stages of the equation:

$$x = 464 \qquad r = 25\%$$

Unfortunately, in this particular format, the data does not work. In the equation, the highly abstract population is expressed as a fraction between 0 and 1. Within this abstract population, 0 represents total extinction, 1 represents the maximum population possible. Thus, we should never have a population below that of zero, ie. with negative numbers, and we can never push our population growth beyond 1. It is worth noting that the non-linear equation cannot be used when the growth parameter rises above 4.0 (Stewart, 1997). A population growth greater than 40% pushes the data through the arbitrary maximum population barrier of 1, rendering the subsequent results meaningless. We need some way of structuring the data to facilitate analysis.

The way to do this in the case of the seed population is to convert the whole number expression of the population to a decimal expression. We have already noted that, for the application of this equation, we cannot ever have a population of more than 1. Therefore, to move our seed population to the right of the decimal point we must divide it by 1000:

$$464 / 1000 = 0.464$$

So,

$$x = 0.4640$$

and we have taken the first step in translating our data into a usable form. It is important to realise that not all seed population data needs to be divided by 1000. Your seed population needs to be to the right of a decimal point so, divide by whatever it takes to get it there.

The growth rate that was indicated by comparison of the *Year 1* population with the *Year 2* population was 25%. Again, this figure is too large to fit realistically within the analysis demanded by the equation. We cannot have a population growth higher than 9.99 recurring. Any population growth of 10 or above would move our fractional population past 1—the established growth maximum. If we divide the growth factor by 1000 as we did with the seed population, the resultant populations move further and further away from 1 until they dwindle away completely. We need to express our population growth in a way that is large enough to be meaningful but small enough to fit comfortably within the constraints of the equations, and the way to do this is to divide the population growth factor by an arbitrary 10:

$$25 / 10 = 2.5$$

This is the second step. Now, not only are we able to express the seed population '*x*' in a material way but we have also managed to express the rate of growth '*r*' appropriately. We are now ready to start to use the equation to show us how this population is likely to develop, providing all other parameters remain unchanged. This can be easily done, using Gleick's stage-by-stage method, with a basic hand-held calculator (see *Table* 3.2).

Table 3.2: Using the equation

• Identify the starting population – *x*	0.464
• Identify the growth factor – *r*	2.5
• Subtract x from 1	1– 0.464 = 0.536
• Multiply by *x*	0.536 x 0.464 = 0. 2487
• Multiply by *r*	0.2487 x 2.5 = 0.6217
• Repeat the process using the new population as a seed	New Population *x* = 0.6217

You need to repeat this process as many times as it takes before you start to see the pattern of equilibrium. These repetitions are referred to as 'iterations'. If you only have to do a few iterations, say ten or twelve, then it is easy to do on a simple hand held calculator. This is how I got the results presented in *Table* 3.3.

Table 3.3:

	1 – x	Multiply by x	Multiply by r (r = 2.5)	$x_{(next)}$
Stage 01: x = 0.4640	0.5360	0.2487	0.6218	0.6218
Stage 02: x = 0.6218	0.3782	0.2352	0.5879	0.5879
Stage 03: x = 0.5879	0.4121	0.2423	0.6057	0.6057
Stage 04: x = 0.6057	0.3943	0.2388	0.5971	0.5971
Stage 05: x = 0.5971	0.4029	0.2406	0.6014	0.6014
Stage 06: x = 0.6014	0.3986	0.2397	0.5993	0.5993
Stage 07: x = 0.5993	0.4007	0.2401	0.6004	0.6004
Stage 08: x = 0.6004	0.3996	0.2399	0.5998	0.5998
Stage 09: x = 0.5998	0.4002	0.2400	0.6001	0.6001
Stage 10: x = 0.6001	0.3999	0.2400	0.6000	0.6000
Stage 11: x = 0.6000	0.4000	0.2400	0.6000	0.6000
Stage 12: x = 0.6000	0.4000	0.2400	0.6000	0.6000

However, if you are going to need many iterations then it is probably easier to set up a spreadsheet to do your calculations for you. Using something like MS Excel will allow you to take your calculations to hundreds or even thousands of iterations if you want to. To set up an Excel spreadsheet, you need to enter the following formula:

$$X_{(next)} = ((1 - \textit{this year's population}) * \textit{this year's population}) * \textit{rate of growth}$$

Table 3.4 shows the spreadsheet with the formula identified. Columns A–D formed the spreadsheet, the data in the final column indicates the spreadsheet formula that was entered when the spreadsheet was set up. *Table 3.4* shows a step by step representation of the formula:

Table 3.4 – Example of spreadsheet showing formula

	A	B	C	D	Formula
1	Stage 1	This Year's population:	0.2586	0.7414	= 1– C1
2		Rate of Growth :	2	0.1917	= D1*C1
3		Next Year's Population =		0.3835	= D2*C2

If you have never used the Excel package before, I will take you through the process step-by-step; if you know how to set up a spread sheet, you can skip ahead at this point.

Setting up your Chaos spreadsheet shouldn't take you very long. The process is best described as three separate phases.

Phase 1: Open up a new spreadsheet and enter the following:

- **Cell A1**: This is the column in which you will keep track of the iterations of the equation, so enter the title 'Stage 1'
- **Cell B1**: Type in the descriptor 'This year's population'
- **Cell C1**: This is the cell into which your seed population data is entered
- **Cell D1**: This is the first cell that needs a formula entered into it. Ensure that your cursor is in the D1 cell and type = 1 – C1. This formula will subtract the num-

ber in C1 from 1 and show the result in cell D1. This gives us the first part of our equation data

- **Cell B2**: Type in the descriptor 'Rate of Growth'
- **Cell C2**: This is the cell into which you will enter your growth rate figure
- **Cell D2**: Enter the formula = D1 * C1. This formula will multiply the numbers in cell D1 by the numbers in cell C1
- **Cell B3**: Type in the descriptor 'Next year's population'
- **Cell C3**: is left empty
- **Cell D3**: Enter the formula = D2 * C$2. This will multiply the value in cell D2 by the value in cell C2 (growth rate) to automatically give you next year's seed population. The dollar sign $ means that the C2 value is an absolute. This means that when you copy all the cells that contain the formula down the spreadsheet, the formula will always refer to the value in cell C2 when carrying out calculations.

Use the 'Help' facility on Excel if you get stuck. It may be a good idea to look in the help index at the terms 'formula', 'absolute' and 'relative' before you start.

Phase 2: When you have entered all of the formula, it is a good idea to put some arbitrary figures into the appropriate cells (remember do not use figures greater than one for your population or greater than 4 for your growth rate). Then:

- Highlight cells A1 to D4 and copy them to cells A5 to D8
- Click on to Cell A5 and amend the descriptor to read 'Stage 2'
- Click on to cell C5 and amend to read = D3
- Click onto cell C6; it should show whatever number you entered for growth rate. Change this to read = C$2.

Phase 3: **Highlight cells A5 to D8 (In the Excel help index, look up the term 'fill handle' to see how to do this easily) and drag this block of cells down the spread sheet to produce as many iterations as you require. I would recommend thirty iterations in the first instance, although you can have as many as you want.**

When core population data is entered into the spreadsheet, the calculation process takes less than five seconds to take the population to 30 iterations as opposed to 20 minutes for 10 iterations using a calculator. As a validity check, you can carry out the computation on a calculator *before* the data is entered onto the spreadsheet. I would recommend that you try a few iterations on a calculator before you convert to using spreadsheets. Only by working the equation through can you really get to understand it and also, it is immensely satisfying to see the data work out. You might find that the end result data varies between calculator and computer, primarily due to the inconsistencies encountered when 'rounding up' data to the arbitrary four decimal points.

This notion of 'rounding up' data is one of the two important things to remember when undertaking the calculation process. One must make a decision regarding 'rounding up' and 'rounding down'. Often the quantitative data that you get runs to more that four decimal places (remember Lorenz and his weather patterns). Data expression and presentation is easier if you decide to keep to four decimal places so you will have to 'round up' the answer at which you have arrived. If you take a number such as 0.19, it is clear that you will 'round up' to 0.2 in much the same way that a number such as 0.11 will be rounded down to 0.1. However, the difficulty occurs when you arrive at a number such as 0.25. It really doesn't matter if you 'round up' or 'round down', but it is important that you are consistent. If you 'round up' the first time you come across this type of result, you must do it for each subsequent result. Although we have spoken of the concept of 'sensitive dependence upon initial conditions', rounding up or down will not affect the predictive nature of the equation unduly. When I carried out the calculations in *Table 3.3* on a hand held calculator, I had to repeat the process twelve times before

I saw evidence of equilibrium. When I checked my findings on the computer, it also took twelve iterations before it identified equilibrium. It didn't matter whether the computer was rounding up or down, the outcome was the same.

The second important thing to remember is that getting someone to check your findings on a different computer may produce unintentionally misleading data. Stewart (1997) emphasises that, as different computers utilise different internal codes, these being the 'private' code that computers use to represent numbers as opposed to the 'public' one that is displayed upon the screen, identical programmes run upon different makes of computer may produce different results. This potential confounding variable should be made explicit, if the tabular format is adopted for the display of results generated by a computer spreadsheet package.

Data formatting with multiple nested systems

There is no law that says you can only use one nested system when trying to identify Chaos in a specific service. The equation application that is outlined in the previous section works just as easily on multiple sources of data as it does on a single source. The difficulty lies in ensuring that the data is presented in a format that is congruent and comparable with the data of any other nested system. If you are just comparing the number of patients seen with asthma with the number of patients seen with chronic obstructive pulmonary disease (COPD), there is no problem. However, if you decide that the nested systems you want to use are patient visits and formal teaching sessions, that is a different matter. You may carry out fewer formal teaching sessions than you do patient visits, but the teaching may be more time consuming or contribute more to the profile of the service. How can this data be made comparable? The answer is to find a variable that is common to both activities that would allow them to be expressed in the same units.

Is there such a variable? Yes there is—**time**. We have already noted that dynamical systems can be viewed within a chronological continuum. Activities as disparate as patient contact and formal teaching sessions can be evaluated in terms of the amount of time taken to carry them out. Therefore, when using multiple nested systems as sources for data, you will need to format the data so that it can be expressed in specific 'units'. What you call these units is up to you (in the illustrative case study that follows in subsequent chapters, we have called them staff/patient contact units or S/PCUs); the important thing is that they are equally representative of all the nested systems you have chosen to use to illustrate your holon. This may feel fairly reductionist in that it is trying to turn a very complex activity, such as nursing, into a simple category or unit. However, in terms of predicting service developments, congruency is vital and the best that can be achieved must be based upon the objective, rather than the subjective nature of care.

When taking this approach, we must make sure that we have a method of expressing our two or more disparate concepts. If we return to our example of using patient contact and formal teaching contact as our nested systems, it is clear that an hour spent talking with patients does not carry the same workload impact as an hour spent running a formal teaching session (it depends on you which one you find more stressful or hardest work). We need a way of translating one nested system into another. This is going to mean looking at both tasks and asking questions again. The sort of questions that you may want to ask, might include:

1. How many patients do I think I visit in an hour and how can I check on this?
2. How many hours of teaching do I do?
3. How many hours do I spend preparing for each hour I spend in class?
4. How can I check if the amount of time I spend is standard or unusual?
5. How much (if any) research and reading time do I get in order to keep myself up to date?

These are just some examples, and obviously the sorts of questions that you want answering will depend upon the nature of the nested systems you have chosen to use. However, the important thing is to be sure that you consider *every* aspect of the nested systems selected in terms of workload demand.

To illustrate what I mean, the steps that we might use to do this are outlined as a worked example in *Table 3.5*. The nested systems we are going to use are those of patient contact and formal teaching. The data set we develop will, at first, be expressed in working hours.

Table 3.5: Calculating working hours

	Question	Answer
Step One	How many patients do we see in an hour?	Service sees average of 4 patients per hour.
Step Two	How many patients do we see in a year?	Records for previous year indicate we see 1400 patients per year
Step Three	How do we express this as 1400 patients as work hours?	1400/4 = 350 work hours
Step Four	How many hours does the service spend in formal teaching each year?	Service provides 1 study day/month. Each study day is 6 hours long; therefore, service provides 72 hours of teaching per year
Step Five	How much preparation time does each formal teaching session need?	Each hour in the classroom requires 2 hours preparation time. So, we must add 144 hour to our 72 hours of formal teaching time

As you can see this gives us a total of 566 (350 + 72 + 144) working hours spent seeing patients and teaching in a formal setting. However, the equation central to this analysis is a population model. The unit of analysis is patient/staff contacts not working hours. In order to apply the equation to the data collected so far, the data needs to be formatted into staff/patient contact units (S/PCUs). To do this, it is necessary to translate working hours into patient/staff units by reversing the procedure that translated staff/patient units into hours:

$$566 x 4 = 2264$$

Be clear about how this works; if we have found that we spend 566 hours in both of the activities we have chosen to analyse, and we have decided that we see 4 patients in an hour, to find out how many patients we see, we multiply the hours worked by the patients seen. This gives us our core population with which to commence analysis—2264 S/PCU. By formatting the data in this way, we have managed to combine two disparate nursing activities for the purposes of analysis. We can now apply the population equation as before using 2264 as our core population. Don't forget that we must also have identified a growth rate in some way. For the purposes of this example, let us assume a growth rate of 20%, thus:

$$X_{(next)} = rx(1-x)$$

is expressed as

$$X_{(next)} = 2.0 * 0.2264 * (1-0.2264)$$

Remember that the * symbol means multiply by. The equation in brackets should be done as a separate calculation ***first.*** Use a calculator to work out the growth of the system and fill in *Table 3.6a.*

Table 3.6a: Self worked application of the equation

	1 – x	Multiply by x	Multiply by r (r = 2.0)	X (next)
Stage 01: x = 0.2264				
Stage 02: x =				
Stage 03: x =				
Stage 04: x =				
Stage 05: x =				
Stage 06: x =				
Stage 07: x =				
Stage 08: x =				
Stage 09: x =				
Stage 10: x =				
Stage 11: x =				
Stage 12: x =				

If you have entered your numbers correctly, you should end up with something very like *Table 3.6b.*

Table 3.6b: Worked application of the equation

	1 – x	Multiply by x	Multiply by r (r = 2.0)	x (next)
Stage 01: x = 0.2264	0.7736	0.1751	0.3503	0.3503
Stage 02: x = 0.3503	0.6497	0.2276	0.4552	0.4552
Stage 03: x = 0.4552	0.5448	0.2480	0.4960	0.4960
Stage 04: x = 0.4960	0.5040	0.2500	0.5000	0.5000
Stage 05: x = 0.5000	0.5000	0.2500	0.5000	0.5000
Stage 06: x = 0.5000	0.5000	0.2500	0.5000	0.5000
Stage 07: x = 0.5000	0.5000	0.2500	0.5000	0.5000
Stage 08: x = 0.5000	0.5000	0.2500	0.5000	0.5000
Stage 09: x = 0.5000	0.5000	0.2500	0.5000	0.5000
Stage 10: x = 0.5000	0.5000	0.2500	0.5000	0.5000
Stage 11: x = 0.5000	0.5000	0.2500	0.5000	0.5000
Stage 12: x = 0.5000	0.5000	0.2500	0.5000	0.5000

As you can see, by stage 5 the system has settled down and no matter how many iterations you put the equation through, you never get any change. We can plot our results on a graph to show us what the holon will look like. You can produce a graph by transferring the end population of each iteration into a data sheet using the 'insert graph' function in MS word. If we plotted these results on a graph with growth rate along the vertical y axis and stages of calculation along the horizontal x axis, it would look like this:

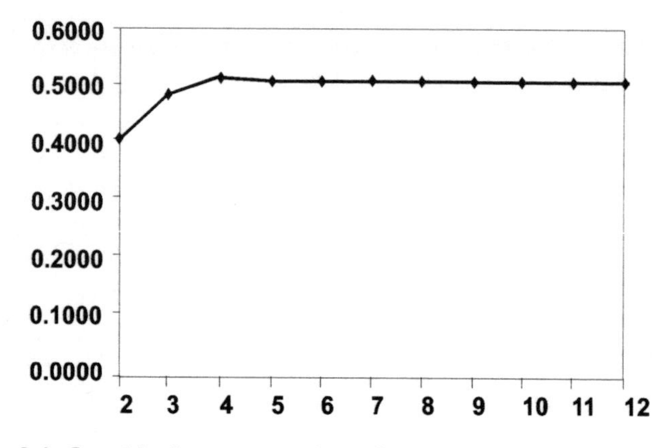

Figure 3.1: Graphical representation of results from *Table 3.6b*

When our original data has been plugged into the population equation and the results have been plotted graphically, we are a step closer to identifying the end state equilibrium of the system. Look at the graph we have generated in *Figure 3.1*. What sort of end state do you think our system is going to end up in? Well, looking at the graph, we can see that by Year 5 the system has settled down and does not show any signs of change, so it is a stable or periodic state. Think back to *Chapter 1* where we found that there are two types of periodic state, which are also referred to as Chaos attractors—fixed point attractors and limit cycle attractors. If you remember Volterra's predator/prey example, you will know that a limit cycle attractor is one that swirls around two fixed points forever (or at least until something happens to change the original scenario), so we would not expect to see a straight line on our graph, we would expect to see a wavy line. Therefore, it is obvious that this service is going to become a limit cycle attractor and rest at one point forever.

What does this mean in terms of how the real service is affected? It means that by the fifth year of development, ***providing all the holon parameters remain the same***, 100% of the service will be accounted for by seeing patients and teaching, because those were the aspects of the service we chose to analyse. This means no professional development, no audit, no meeting attendance, no clinical supervision, etc. Clearly, that is not what most people would want for the service they provide. It must be emphasised that this prediction, like most predictions, is not a necessarily a true representation of the future. What it can do is demonstrate one possible service end state, if nothing is done to change the overall trajectory of the service. Its use lies in allowing for forward planning in terms of staff recruitment, review of service parameters or restriction of service growth. If you want to, you can change the system parameter to see what effect this may have on the service you provide. This notion of parameter manipulation and end state prediction will be explored in the following chapters.

Summary

This chapter has introduced the concept of the population equation that will form the foundation of all of your future calculations. It has also spent some time explaining the stage by stage application of this equation. The notion of identification of holon parameters has been explored and the importance of data formatting, if you choose to explore more than one aspect of your practice has been emphasised. We have started to explore the concept of mapping the results of the population equation, in order to identify what type of chaotic attractor our service might reasonably evolve into, ***providing all the holon parameters remain the same***—a very important point. This prediction can be used to analyse service need in terms of extra resources that might be required. Predictions can be manipulated by changing certain holon parameters and this will be clarified later. The next chapter will introduce you to the case study that will form the basis of our subsequent discussions regarding the application of Chaos Theory.

4

Using Chaos Theory: An illustrative case study

The best way to come to terms with a technique as hugely abstract as this one has been is to see it work in practice. To this end, this chapter will take the form of a case study that will evaluate the collection and application of data from an acute pain service, which will be referred to as APS2. The identification of holons and holonic parameters, the manipulation of data into a usable and meaningful format, and a further stage-by-stage application of the population equation that you should, by now, be familiar with, will be described, together with the rationale for such manipulation. Where appropriate, the lessons learned from previous applications of this system will be highlighted.

The nonlinear population equation that is fundamental to this work could, in reality, be applied to any service, specialist or otherwise. It is as appropriate to use it to analyse ward-based services as it is to use it in clerical or educational situations. However, for the purposes of this case study, we will use an acute pain service for two main reasons. Firstly, pain services have been established in the UK for well over ten years, since the publication of the joint colleges' working party report into pain after surgery (Royal College of Surgeons and College of Anaesthetists, 1990). It was this report, and the importance it attached to the development of acute pain teams, that laid the foundation of acute pain management today. Thus, it can be seen that acute pain teams are among the longest established of all specialist teams in UK nursing. This means that their systems of work and occupational boundaries are well established and, as such, are comparable with health care that is rooted in settings of a more generic nature. This links with the other important point that dictated the use of acute pain teams for this study. Both the Audit Commission

(1998) and the Pain Society (1997) have highlighted the importance of audit and evaluation. The Pain Society suggests that the management of pain is:

> *'...important in terms of comparing institutions delivering similar health care. Measures* [of efficiency] *will assess staffing ratios, case mix.....and patient turnover'.*
>
> The Pain Society (1997: 14)

This implies a strong focus on the keeping of *patient contact statistics*, which is of great importance in a study of this nature. As we have seen, an understanding of the nature of the work to be evaluated is vital, and this understanding is facilitated when statistics and work-based data are readily available. The type of service analysis that was carried out for this case study might be useful to you when you start to look at your own practice.

Identifying holons and setting parameters

The service used in this case study is based in a district general hospital with around 800 beds on its main site, serving both rural and urban communities. The hospital has an acute pain service that has been in place for eight years. The pain service consists of one acute pain nurse and one consultant anaesthetist. The service provides a nine-to-five, five-day-per-week service, and is supported at ward level by trained pain link nurses.

A local pain research group was approached to take part in a group interview in order to attempt to gain a consensus statement on the basic nature of the work of the acute pain practitioner. Although the nonlinear equation being used can be applied using only the most basic data, in order to ensure validity and reliability, an in depth analysis of the boundaries of service provision should be undertaken. Previous experience has shown that, if the boundaries of provision are not fully considered at the outset, the chaotic data obtained cannot be conceptualised in any useful way. We have already noted that one of the fundamental elements of Chaos is articulation of the nature of a dynamical system and that this articulation must utilise qualitative methodologies, such as focus groups, to obtain data that is qualitative in nature but amenable to quantitative expression, with non-participant observation as a way of testing construct validity. This multi-paradigm or multi-method approach to information-gathering is one that is not always seen as 'pure' when undertaking research projects, but is appropriate in this case, since the numerical data we will be using must be firmly based upon the experience of the service providers. The focus group consisted of eight acute pain practitioners from different national health service hospital trusts within a one hundred mile radius of the APS2 hospital.

In the first instance, the group was asked what elements of its role was considered to be core to the service provided as acute pain practitioners. Overwhelmingly, the response was patient contact and teaching. The group was then asked to calculate *on average:*

- How many patients were seen in one hour
- How much formal (i.e. not bedside) teaching was carried out in a month
- How much preparation time was allowed by the group when teaching in formal settings.

A note of caution must be sounded regarding the make-up of any focus groups that you might want to use. This group was quite diverse in that each group member worked in very different settings, from small local general hospitals, to medium size trusts and large teaching hospitals. Depending upon the nature of the service provided, taking an average from a diverse group may compromise your data. In this case, group consensus was that the locale was incidental and that the demands of the patients and the interpretation of the role of the acute pain practitioner were largely the same. In response to number of patients seen, the group felt that three patients per hour was an appropriate representation of the member's work load. On the subject of teaching, the group could not reach a consensus. In the field of education, size of the hospital was a factor; practitioners in large teaching hospitals did more formal, classroom-based teaching than colleagues in smaller hospitals, although both sets of practitioners did teach the same groups of students, predominantly doctors and nurses. However, following much group discussion and role analysis, the consensus of the group was that three hours preparation time (preparation time being reading, preparing teaching materials, arranging study sessions, liaising with external speakers, etc.) for every hour of classroom contact was a reasonably accurate reflection of the work involved.

The data obtained from the focus group indicated that a pain practitioner, covering ward-based populations, would see an average of three patients per hour. It was also agreed that, although the number of formal teaching sessions varied, preparation time of three hours for each hour of formal teaching was accurate. However, as a way of ensuring the construct validity of the focus group as a measurement instrument, it was appropriate to undertake non-participant observation of two of the eight practitioners in the clinical setting.

Construct validity should be examined as a method of testing theoretical assumptions (Oppenheim, 1992), so in order to test the validity of the data generated by the focus group, it was agreed that a 'typical day' should be witnessed via non-participant observation. Two volunteers from the original focus group were used as the sample for this validity check. Permission to accompany the practitioners throughout their working day was obtained from line managers. There was no necessity for hospital ethical committee involvement, as patients would not be treated or have their care influenced or manipulated by the observer in any way. All of the patients visited with the pain nurse were told why another person was present and given the opportunity to ask them to leave. No patients availed themselves of this opportunity, although, had they done so, it would not necessarily have compromised the data, as the focus was upon the time spent with the patients rather than with any insights into practitioner/patient communication. Each of the two pain practitioners was accompanied throughout, what they both described as, a 'typical sort of day'. It was interesting to note that the patients per hour agreed by the focus group appeared to be accurate when averaged out over the number of patient contact hours in the working day.

There are two important points to be emphasised here. Firstly, a full picture of practitioner activity cannot necessarily be extrapolated from one day's observation. If you choose to utilise this approach, you will need to negotiate a much longer period of observation. Secondly, it is not compulsory to use focus groups for functional analysis. This is just one method and you may wish to explore other options when you start to examine your own service. When you start to evaluate your practice for yourself, you may decide that setting up and running a focus group is too difficult or too time consuming. There is no right or wrong way to go about this. The most important thing is that you are very clear about which aspects of your practice you are going to focus upon, and how much time you devote to them. You can carry this review out on your own or you could get someone to help you by shadowing you and keeping notes, although this is time consuming and labour intensive because it should be done over a period of several days or, preferably, weeks if a comprehensive picture is to be obtained.

The data obtained from the focus group and the non-participant observation provided insight into the nature of the dynamical system that was the acute pain service, and allowed for the identification of the nested systems within the holon. In this case, our holon was the pain service itself and the nested systems we were looking at were patient contact and staff education. We knew that when we spoke of this specific acute pain service, we were really considering the educative and patient contact aspects of the service with all other activity being discounted as 'noise'.

Data collection and formatting

The nurse practitioner of the acute pain service supplied the patient contact data that would allow us to calculate service growth rate. In order to ensure that there was no generational overlap as outlined by May (1976), data was compared from the previous two years of the service. The APS2 service had seen 1357 patients in 1998 (January–December) and 1697 patients in 1999 (January–December). This shows a growth of 20% over twelve months; therefore 20% formed the basis of our growth rate. In a previous pilot study, we had concentrated upon one treatment modality only, that of patient controlled analgesia (PCA), to the detriment of our end data. We had found that, by only concentrating on one tiny sub-set of patients (those with PCA), we lost a great deal of other patient contacts that should, on reflection, have been included in our calculation. This would suggest that taking a sub-set of patients weakens the idea of nested systems to the point that it becomes meaningless and no overview of the service can be obtained. In this case, these figures included *all* patients that were seen by the pain practitioner during this period. This provided a more complete picture of the demands of the service within the context of that particular nested system. The non-participant, observation-validated data generated by the focus group allowed us to identify how many patient contact hours the practitioner had used while seeing these patients. Focus was upon the 1999 data. Consensus opinion of the focus group was that one pain practitioner could see approximately three patients per hour. Therefore, to translate those patient contacts into service hours:

$$1697 / 3 = 565 \textit{ hours of work}$$

The focus group had also suggested that formal teaching time should be incorporated into the equation, arguing that a model that suggested three hours of preparation time for each hour of formal teaching contact was appropriate. The APS2 pain service had recorded 74 hours of classroom contact in the 1999 period. So, to format this data:

$$74 + (74 \times 3) = 296 \textit{ hours}$$

and to ascertain actual contact hours, which include patient and staff teaching in formal settings:

$$565 + 296 = 862 \textit{ hours}$$

It became apparent that the pain practitioner of this service spent approximately half their time in clinical or educational contact. This figure is arrived at thus:

$$566=296=862 \ (\textit{Actual staff / patient contact})$$

$$37.5 \times 52=1950 \ (\textit{Working hours in one year})$$

$$37.5 \times 7=263 \ (\textit{Seven weeks annual leave})$$

$$1950-263=1687 \ (\textit{Actual contracted hours})$$

$$1687 / 862=195 \ (\textit{Half contracted hours})$$

However, as we are aware, the equation central to this analysis is a population model. We have chosen staff/patient contacts not working hours as our unit of analysis. Therefore, our data needs to be formatted into staff/patient contact units (S/PCU) by translating working hours into patient/staff units. We do this by reversing the mathematical procedures that translated staff/patient units into hours:

$$862 \times 3=2586$$

This gives us our core population with which to commence analysis—2586 S/PCU. From now on, we will refer to the core population as the seed population because that is the population value that we will use to seed our holon growth and development. Remember the S/PCU units of measurement are relatively arbitrary. They do not 'measure' anything, they are simply a convenient way of marrying two sets of disparate data together for analytical purposes. When you start to look at your own practice, you can develop descriptive units that best fit the elements of the holon that you are going to plug into the equation.

It must be emphasised that this method of data formatting can be criticised for what it omits as much as for what it includes. It is obvious that no pain service spends all its time in this fashion. Multidisciplinary and cross boundary working, record keeping, professional development and participating in research projects and audit have all been identified as part of the role of the pain practitioner (Royal College of Surgeons and College of Anaesthetists, 1990; The Pain Society, 1997; The Audit Commission, 1998). Nevertheless, such elements can be identified as 'noise', outside data that may affect the analysis of Chaos within a system (Rapp, 1993). Such noise should be acknowledged and, if possible, set aside as deflecting attention from the specific data under consideration. If you want to, it is possible to include these elements, formatting them in much the same way the formal teaching data was formatted. However, May (1976) has warned that too many data elements render the identification of Chaos in a system less transparent. It is almost as if you get 'Chaos overload' if you try to look at too many nested systems at once. This can be quite tricky; generally, when undertaking a research project, the more data you have, the more rigorous your findings are seen to be. But this is not research, neither is it audit. It is simply a method of extrapolating service development over time. This feeling that the more nested systems the stronger the data, is one that should be guarded against. This is the 'rock and a hard place' perceived by Rapp (1993), in that numerical tests for Chaos require large amounts of data, but over sampling biases the results. Looked at logically, if you included all of the nested systems that contribute to your holon in the time dependant equation, the results would be meaningless since including everything you do would account for 100% of the service. It must also be highlighted that, for some aspects of your service, data is both specific and sparse, but this may not necessarily pose the same problems that small data sets cause in research studies. At all times, it must be remembered that this process provides a time dependant snap shot of your service; it is not necessarily specific to your entire role, neither is it predictive. It can only show you what will happen if all other parameters are unchanged.

This brings into prominence, the difficulties that were predicted by Marks (1994) in terms of the methodological challenges facing the researcher attempting to apply a predominantly 'scientific' approach to a multivariate problem. This means that seeking to define a cause and effect type of response may not be appropriate, particularly when dealing with many different aspects of the same service. However, it is contended that, by basing decisions on data inclusion on the experience of the service providers, the end result will still be meaningful to that specific service. This approach also conforms with the proposal from Rapp (1993) that suggests data is more meaningful when sampling criteria are in place. He suggests those sampling criteria, as used here, help to reduce noise contamination So, if the main focus of a service is upon seeing patients, for example, any other aspect of that service should be categorised as noise and disregarded for the purpose of Chaos calculation. This means that, when you have identified your holon and the nested systems within it, you should select just one or two to form your sampling framework.

Within this case study and subsequent to data collection and formatting, the seed population was identified as 2586 S/PCU and our growth factor was 20%. However, experience from early studies has clearly indicated that whole numbers such as these are inappropriate for use with the equation:

$$x_{(next)} = rx(1 - x)$$

The numerical data obtained thus far must be further formatted for use as a fraction in a highly abstract population that falls between 1 and 0. The importance of maintaining quantitative integrity when formatting the seed population had been emphasised; therefore, the seed population was expressed as:

$$x = 0.2586$$

Like the units of measurement (S/PCU), the expression of the seed population is arbitrary. What I mean by this is you would not always divide your whole number population by a factor of 10,000 as we have here. The important thing is that you express a whole number population as a fraction. If our seed population was 258 we would divide by 1000, if it was 28 by 100. The population growth factor that had been identified as 20% when the baseline data was provided was likewise formatted for use with the equation. The end result was that the baseline data was expressed so

$$x = 0.2586 \qquad r = 2.0$$

Data analysis

When the seed population data was entered into the spreadsheet, the calculation process took less than five seconds to take the population to 30 iterations or repetitions as opposed to 20 minutes for 10 iterations using a calculator. As a validity check, the computation was carried out on the calculator *before* the data was entered onto the spreadsheet. It was interesting to note that, although the data varied between calculator and computer, primarily due to the inconsistencies encountered when 'rounding up' data to an arbitrary four decimal points, by stage four both techniques agreed on the numbers and the end result was identical (see *Table 4.1).*

Table 4.1: Results for data from APS2

	1 – x	Multiply by x	Multiply by r	x (next)
Stage 01: x = 0.2586	0.7414	0.1917	0.3835	0.3835
Stage 02: x = 0.3835	0.6165	0.2364	0.4728	0.4728
Stage 03: x = 0.4728	0.5272	0.2493	0.4985	0.4985
Stage 04: x = 0.4985	0.5015	0.2500	0.5000	0.5000
Stage 05: x = 0.5000	0.5015	0.2500	0.5000	0.5000
Stage 06: x = 0.5000	0.5015	0.2500	0.5000	0.5000

When a system shows clear equilibrium as the APS2 system does, the graphical representation is unambiguous (see *Figure 4.1*).

Figure 4.1 shows the plotting of the data to 20 iterations; however, the trajectory of the system is overt. It indicates that, if the acute pain service of the APS2 trust continues to operate with its current growth parameter of 20%, the service can only hope to persist for four years. After that point, the entire focus

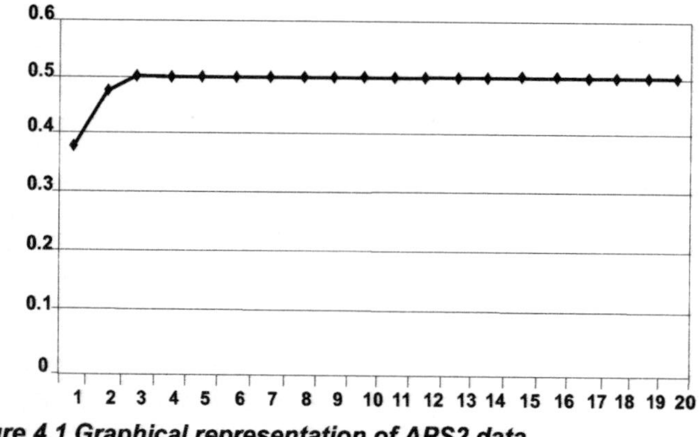

Figure 4.1 Graphical representation of APS2 data

of the service will be upon patient contacts and education with absolutely no time for any other tasks. This is well demonstrated by re-formatting the data from S/PCU to hours. The stage 5 seed population is expressed as $x = 0.5000$ (***Table 4.1***). If this factor is removed from the arbitrary parameters of the equation and articulated as non-decimal SPCU, it equates to a population of 5000. If the data formatting described earlier in this chapter is reversed in order to translate S/PCU back into practitioner hours the following result is achieved:

$$5000 / 3 = 1667$$

This equates to 99% of a pain practitioner's contracted hours

State attractor identification—APS2

It can be seen from the data presented that the APS2 pain service is a system that will reach stable equilibrium within four years. The service will continue to grow and develop up to that point, but afterwards it will remain unchanged over time unless the parameters are changed or some outside intervention alters the service. To express this formally, *the APS2 pain system is a fixed point attractor.* Now that the end state has been identified, we can examine the service to see what we can do to avoid this outcome. The important point to remember is that this is what will happen to the service if all of the holon parameters remain unchanged. The questions that this exercise raises may include:

1. Is a four year life span all we want from this service?
2. How likely is it that the holon parameters will change?
3. Do we want to extend the life of this service?
4. Do we want to actively change the parameters to affect this indicative outcome?
5. What will happen if we do manipulate the parameters?

Summary

This chapter has examined data from a specific patient service. The data was subjected to explicit transformation, which rendered it compatible with the population equation that is the foundation of Chaos hunting. It is argued that this formatting did not detract from the power and validity of the data since, although the different elements of the data were rendered homogenous and expressed in terms of a similar measurement term, the meaning of the original data remained intact. The development of a spreadsheet formula has been outlined and the speed and efficiency of this method of data analysis has been emphasised. The data has shown some predictive power in that it has been possible to identify the APS2 pain service as a fixed point attractor that will reach an equilibrium state within four years of the original data collection. This forecast assumes that the growth parameter that was the foundation of this calculation will remain unchanged. The questions this exercise raises have been outlined and the option of deliberate parameter manipulation has been introduced. The following chapter will consider the effects of parameter manipulation of the attractor state of the system.

5

Parameter manipulation

So far it has been demonstrated that, provided sufficiently rich data is available, it is possible to determine the end state of a given dynamical system. In *Chapter 4*, we reviewed how this works in practice by using the APS2 case study data to calculate the life expectancy and end state of specific nested systems within a holon. We looked at the type of questions we should be asking ourselves when undertaking the sort of service analysis outlined by the case study. Two of the queries were:

1. Do we want to actively change the parameters to affect this indicative outcome?

2. What will happen if we do manipulate the parameters?

The purpose of this chapter will be to examine the answers to those questions by investigating the outcomes that can be obtained when one system parameter, that of population growth is manipulated. This means that we will use the data from APS2, as described in the last chapter, and manipulate the system boundaries, specifically the parameter of growth rate, to see what will happen. The seed population from the APS2 data will remain unchanged at 0.2586 S/PCU; however, we will change the system parameters from a growth rate of 10% per annum, expressed as 1.0, to a growth rate of 40% (4.0) in 5% stages. Each specific attractor state of each stage will be identified and the concluding section of the chapter will consider the different attractor states exhibited by the data. This should be useful to you when you start modelling your own system outcomes, because it will enable you to picture what your system end state will be.

Decreasing the growth parameter

The growth parameter of 20% that was applied to the original APS2 data in the previous chapter was based upon service audit and suggested an equilibrium point for the service within four years. If the same seed population is arbitrarily subjected to a reduced growth rate of 10% year on year, the system shows the development pattern as represented in ***Figure 5.1.*** This figure shows that, when the service is growing at a rate of 10% year on year, the service is sliding towards extinction with numbers declining each year. It is worth bearing mind, at this point, that this does not mean that there are less patients in pain within the hospital, only that less patients are being seen by the acute pain service. This may be the situation if a service decides it will limit the patients it sees in some way, only seeing those patients in care directorates that contribute money to the service, for example, or not visiting patients in other specialities. The decline begins quite sharply, but gradually starts to develop a gentler slope. At first glance, it looks as if it is evident that the system is evolving into a fixed point attractor and will continue to decline until it reaches extinction. However, if you take the iterations to many places you will find that this is not really the case.

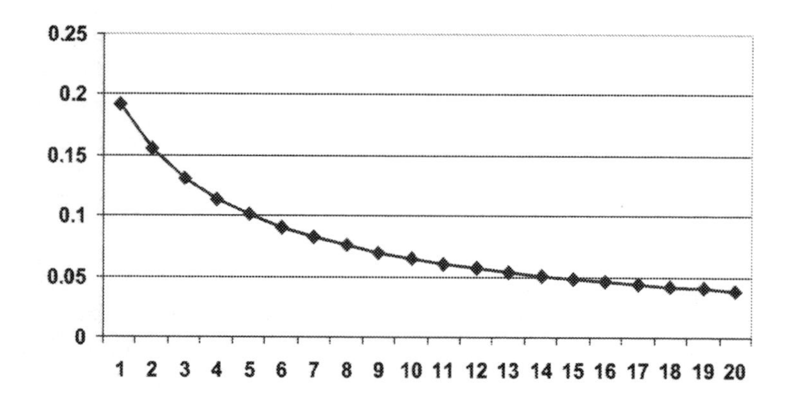

Figure 5.1: – r = 1.0

This graph shows 20 iterations or repetitions. When it became clear that a growth parameter of r = 1.0 was suggesting system extinction, the spreadsheet iterations were extended in order to try to see exactly how long it would take for the system to disappear. This is not always easy to do, so, for this example, the results were more than a little unexpected. The calculation was taken to over 1100 iterations (this is one of the great things about using the spreadsheet package—imagine how long 1100 iterations would take by hand) and the system was still demonstrating some, albeit very small, evidence of population. By iteration 1107 the seed population was expressed as x = 0.0009. It should be remembered that the use of four decimal places was an arbitrary one based upon the examples given by Gleick (1987) whose model of population calculation was utilised within this study. If we had insisted on rounding our calculation up to only three decimal points, the system would have experienced extinction by iteration 1042—still a very long time. In reality, of course the service would have expired long before that time because it was inefficient and not cost effective. A service that sees less people each year cannot be seen as proficient except in scenarios where one would expect patient demand for the service to fall in a corresponding manner.

The behaviour of the quantitative data was also of some interest. As the system progressed toward extinction, it demonstrated periods of equilibrium, small periods when the data seemed to follow a stable and reasonably linear trajectory, for example, by iteration 163 x = 0.0058. This period of stability last for a further eight iterations before the system recommenced its decline. This would seem to be a manifestation of the 'periodic outbreaks' identified by May (1976). May found that, when he applied this population equation to biological populations, growth (or decline) would continue and then the system would appear to rest in equilibrium for a short space of time, usually, before splitting off into a different pathway of growth. On the other hand, it must be emphasised that this might simply be due to the vagaries of the computer system used, and should be treated with caution as not really being of practical significance. It is presented here simply as a matter of interest. Ideally, the data should be checked and validated on different hardware before any true pattern can be detected. The factor of change between each seed population also demonstrated these periodic outbreaks of equilibrium. The period of change likewise slowed as extinction point was reached. As one would expect, the period of change was non linear in nature.

So what does this mean to your service? It means that, if you decide to change the focus of the number of patients seen, certainly the longevity of the service will be enhanced, but it will slowly decline year by year. Why would anybody wish to do this? It could be that one of the goals of the service is the transfer of skills to ward-based clinicians so one would expect that the demands upon the service would decline annually. This type of parameter manipulation would help you to identify how long it might take for the service to become redundant. One the other hand, it may be that service levels have been set too low and this type of manipulation will highlight that, allowing for the number of patients the service is willing to carry to be increased in an informed manner.

A very different picture occurs when the system growth parameter is set at 15% or 1.5.

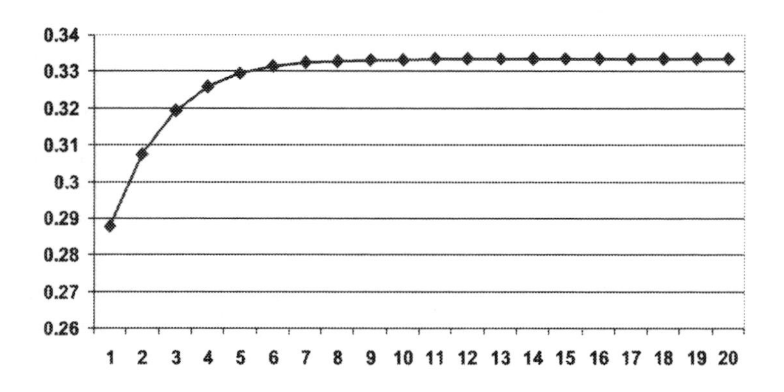

Figure 5.2: r = 1.5

Far from sliding towards eventual extinction, the dynamical system shows a very similar pattern to that demonstrated by our original mapping exercise in *Chapter 4—Figure 4.1*. The fundamental difference is that not only does the system with a growth parameter of 1.5 take longer to reach an equilibrium point, i.e. 11 iterations as opposed to the 4 shown in *Figure 4.1*, the system also settles to its fixed point attractor state at a lower level; 0.3333 rather than the 0.5000 demonstrated by the original APS2 calculation. Once again the system is manifesting the characteristics of a fixed point attractor.

In terms of service delivery, this means that, as before, you can track the end point of your service. However, if you wish to set your growth rate slightly lower—15% rather than 20%—you can allow your service to develop at a slower rate. This might be useful to you if you are expecting to recruit new staff members who may need to attend training programmes or if you are setting up a new service and feel that there might be a high demand that you would like to control.

Increasing the growth parameter

So far it can be argued that the APS2 system has been shown to be a reasonably stable fixed point attractor. Regardless of the 'r' value utilised, the system eventually settles in a stable equilibrium state, as a fixed point attractor, or as near to that state as makes little difference. This state of affairs continues when the growth parameter is raised by 5% from the original growth factor of 20% so that r = 2.5.

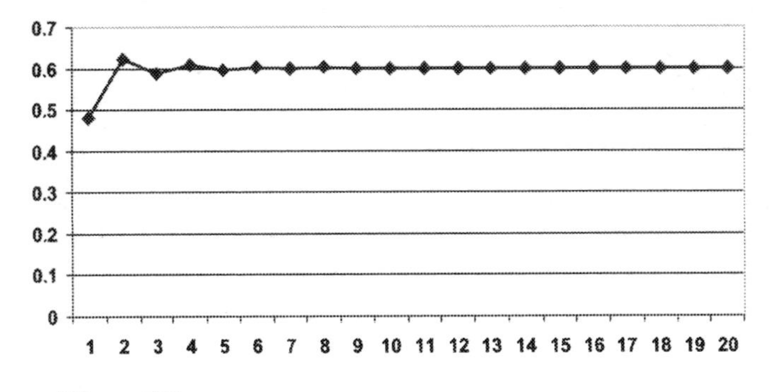

Figure 5.3: r = 2.5

Figure 5.3 shows the system prediction for an annual growth rate of 25%. Once again it can be seen that the system is a fixed point attractor and shows a similar response curve to the parameters, r = 1.5 (*Figure 5.2*) and r = 2 (*Figure 4.1*). It is noteworthy that the equilibrium point still occurs at iteration 11, but at a level of approximately twice that demonstrated when r = 1.5, 0.6000 as opposed to 0.3333. This would suggest that, in this particular equilibrium system, a 10% rise in growth rate manifests itself as a doubling of the end point population. However, it is not possible to extrapolate this to other equilibrium systems without further data collection and analysis; you would need to try it out on your own service holon to see if this is the case with your data.

This equilibrium state becomes threatened as soon as the growth rate parameter is changed to r = 3.0 (see *Figure 5.4*).

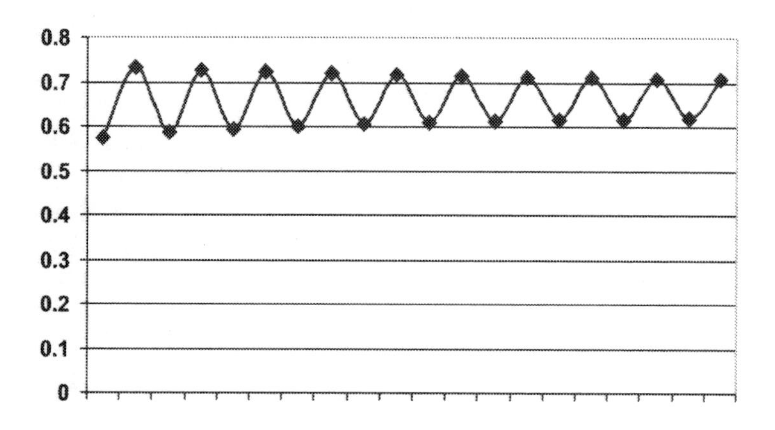

Figure 5.4. r = 3

The dynamical system that was so stable when growth parameters were restricted to between 1.5 and 2.5 starts to develop chaotic characteristics when the growth rate is pushed to an annual increase of 30%. *Table 5.1* clarifies the oscillations of the numerical data.

Table 5.1: Data oscillations for r = 3

Iteration	x =
1.	0.5752
2.	0.733
3.	0.5871
4.	0.7273
5.	0.5951
6.	0.7229
7.	0.601
8.	0.7194
9.	0.6056
10.	0.7166
11.	0.6093
12.	0.7142
13.	0.6124
14.	0.7121
15.	0.615
16.	0.7103
17.	0.6173
18.	0.7087
19.	0.6193
20.	0.7073

Thanks to the fact that iteration is so easy to carry out with computer technology, you can take this to as many repetitions as you wish. I took this calculation to 120 iterations with no sign of equilibrium. The system swirls round and round the data set outlined above. It is accurate to state, therefore, that with a growth factor of 30% our dynamical system becomes a limit cycle attractor.

In practical terms, we have probably all been in situations when we try so hard to meet the demands made upon us, and for some time we manage. Then it all gets a bit too much and we become less efficient. However, after a period we feel equal to meeting the service demands again and the cycle starts over. Another way to look at this might be to consider your behaviour when you start a programme of study. At first you are very enthusiastic and quite able to keep up with the course work, then as the course gets harder and it becomes more difficult to juggle study, home and work, you find that you are less able to devote time to your course work. After a while though your conscience may prick you and you return to your course work again.

One thing to be aware of when looking at *Figure 5.4* is that the limit cycle attractor state looks fairly predictable, with the peaks and troughs reasonably evenly spaced. The true elements of chaos factors become overt when the growth factor parameter is raised to 35% (*Figure 5.5*) .

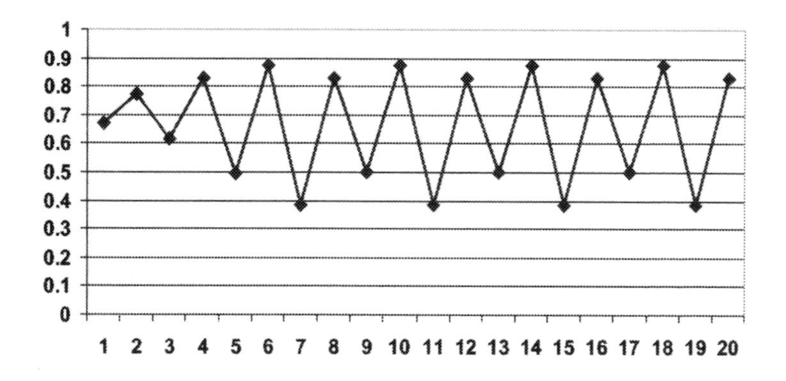

Figure 5.5: r = 3.5

Note how the oscillations become less regimented as the system becomes more chaotic. These fluctuations are a constant feature of this parameter manipulation, even when the calculation is carried out to 500 iterations. They become even more overt as the parameter is increased. *Figure 5.6* shows the system trajectory for growth parameters of 4.0. It is worth noting that the non-linear equation, which has been utilised throughout this study, cannot be used when the growth parameter rises above 4.0 (Stewart, 1997). A population growth greater than 40% pushes the data through the arbitrary maximum population barrier of 1 rendering the subsequent results meaningless. Try it for yourself and you will see.

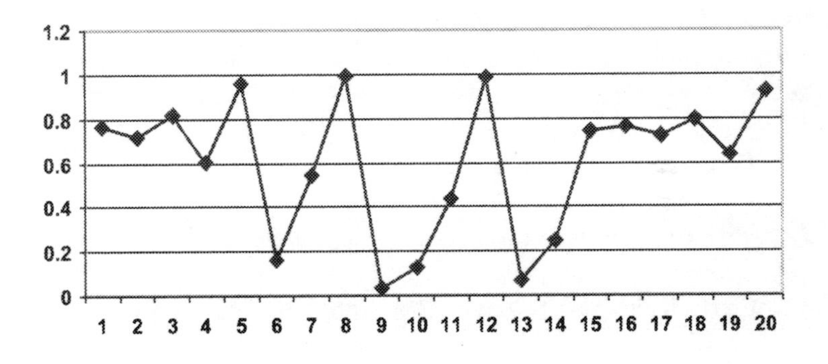

Figure 5.6: r = 4.0

It is interesting to note that the stable period cycle referred to by May (1976) can be tentatively identified at iterations 1–4 and 15–16 of the r = 4 data.

In terms of your service, any holon that demonstrates this as an end state attractor is really in big trouble. It is a strange, or Chaos, attractor. The service is being expected to develop far too quickly with demand outstripping resources. There is no way to forward plan and the entire service will be in disarray. This is not a good picture to have when you plot your service out.

Why would you bother to manipulate your system parameters in this way? After all you have already plotted the development of your service and probably managed to identify what sort of end state attractor it is likely to become. We have already established that applying the population equation to your service will only give you a brief overview of one potential end point, an end point that is only valid if all the original parameters remain unchanged. Nevertheless, you might want to know what will happen if things change, perhaps you will employ more staff or reduce the number of referrals your service will take, thereby lowering the growth rate of your service. Experience suggests that when you are manipulating parameters to model outcomes of delivery change, it is best to alter your growth rate by factors of 0.5; i.e. 1, 1.5, 2, 2.5 and so on. If you alter it by a smaller amount, the change in end state might not be clear enough; if you alter it by too much, you might miss an important attractor state. The sorts of questions that you would want to ask yourself might be:

1. What happens if I reduce my growth rate?
2. What happens if I increase the growth rate?
3. Can I recognise the attractor state that I have plotted?
4. Can I successfully transform the data into working hours?
5. Can I express the data as a percentage of the total service delivery?
6. Can I use this data to forward plan the development of my service?

Identifying attractors

This chapter has focussed exclusively on the trajectory changes that are evident in the APS2 seed population of 0.2586 S/PCU when the growth parameters are altered. It can be seen that the optimum growth rate for this dynamical system is one where r = 2.5. In this instance, the service reaches an equilibrium point at a later stage than previous calculations, giving a slightly longer life span. It can be postulated that all the data that was within the parameters r =1.5 to r = 2.5 can be classified as a fixed point attractor. With the exception of the r = 1 data set, all the iterations increased steadily until they maintained an equilibrium level. In terms of specific aspects of a service, this means that the service would grow at a regular rate for varying numbers of years (or months or weeks, whatever the time span you choose to use) until it could not grow any more, and 100% of the service time was taken up with whatever aspect of your service had formed the focus of your calculation.

The r = 1 data set was slightly different in that this was the only data set which showed how a service could slide towards eventual oblivion. However, it could be argued that the r = 1 data set demonstrated a 'negative' equilibrium; it never became wholly extinct and demonstrated increasing long periods of equilibrium. This shows that for a full time service to be considered viable, it should be able to guarantee a more than 10% year on year growth. Less than 10% growth may be acceptable if the service is only being provided on a part-time basis, but this requires further exploration.

The pivotal data set, for this assemblage of service data, appears to be the r = 3 data. *Figure 5.4* plots the iterations as a wave like representation with the system oscillating between two main data points. Subsequent repetitions of the calculation indicate that the system will continue in this manner indefinitely. The development of this data set is congruent to that of the Volterra predator/prey population model (Kosko, 1993; Cohen and Stewart, 1994). The system will continue in this 'boom and bust' cycle until some outside force manoeuvres it into a new state. In service delivery terms, this end state represents a facility that quickly becomes strained beyond its ability to meet demand and responds by cutting back its service provision. As provision is reduced and demand drops, the resources once more become available, demand upon the service rises once again and the cycle begins again. This is a situation that I am sure most of us have experienced, but if it is identified early enough in the life of a service, then staffing recruitment or referral rates can be planned for.

On the other side of the r = 3 limit cycle attractor, we can identify explicit and overt chaotic attractors. The diagrams *Figure 5.5* and *Figure 5.6* clearly indicate that the system has reached a state of aperiodic equilibrium or Chaos. The data fluctuates wildly and with no apparent order. This is indicative of a service that is really out of control, in which every day sees the staff 'fire fighting'—i.e., responding with a knee jerk reaction to the demands made upon the service rather than attempting to manage provision of care in an ordered way. This too is a state I suspect most of us are familiar with; the situation occurs where you are so busy that you start off by trying to prioritise the care given and very quickly reach a state where everything is of equal priority and all coping mechanisms begin to collapse.

The interesting thing about this parameter manipulation is that, no matter what your original seed population, the growth rates outlined here will generally exhibit the same end attractor state. Thus, a growth rate of 3.5 will usually demonstrate Chaos regardless of the original population number plugged into the equation. The only thing that differs is the amount of iterations it will take you to reach whatever attractor states the nested systems will settle into. Therefore, you could assume that setting a growth rate of 20% per annum for your service will ensure that it will become a fixed point attractor, but you will not know for sure and you will not know how long it will take to achieve that state. You will only find that out by plugging your data into the equation and letting it run.

Summary

This chapter has manipulated the parameter of growth rate from the nonlinear equation that has been applied to the APS2 data set. The various graphical representations that have been developed allow the topology of the system attractor state to be plotted and identified. This approach mirrors the work of Martinerie *et al* (1998) who utilised EEG recording to map attractor topology and identify Chaos in the epileptic patient. It has been noted that the data set utilised for these manipulations suggested that any growth rate of less than $r = 3$ was a limit cycle attractor. It was contended that the $r= 1$ data inferred some kind of negative equilibrium in that, even though the system was sliding towards extinction, there were considerable periods of equilibrium identifiable within the trajectory. This resonated with the work of May (1976) who previously noted such intervals. However, while this is of added interest, it does not really affect overall planning of service development which is, of course, the main focus of the application of the equation in this work, at least.

Once the 'r' parameter is moved above 3, the system begins to glide towards a chaotic state. The fluctuations of the seed populations can be seen as a manifestation of the chaotic attractor that is characteristic of this state. Thus, it can be seen that consideration of the manipulation of the population parameter within a nonlinear equation can provide insights into the various attractor states of the seed population. We have emphasised that the end attractor states for growth factors will consistently be the same, but the amount of time taken to settle into that state is sensitive to the quantitative data that is plugged into the equation. The next chapter will review the application of the population equation this far, and then consider how this information can be used when forward planning for specialist care services.

6

Utilising nonlinear information

So far, this book has demonstrated that the application of a nonlinear modified Malthusian equation can provide insights into the development any service (the example we have used so far is that of a specialist pain service) provided that sufficiently rich data is available. It may be prudent at this point to review the entire process again.

Stage 1: Describe your holon

Remember that in *Chapter 1* we introduced the concept of a holon as a convenient way of referring to gestalt systems. By this we meant systems that are understood to have many systemic components, but that are viewed as whole systems for the sake of convenience. So, we use the term 'holon' to describe the arena of care in which we function. A ward, a specialist service or even a shift can be seen as a holon. To refresh your memory, look back at *Table 1.2* in *Chapter 1*, which summarises the important concepts surrounding holon recognition.

Stage 2: Identify the nested systems

Within the holon you have described, you should be able to distinguish one or two areas of practice that, taken in isolation, still give a reasonable idea of the nature of the holon. These are your nested systems and are usually the focus of the calculation of Chaos. Nested systems may be as diverse as seeing patients in a specialist service, dealing with members of the public in reception areas, or teaching students in formal or informal settings. Do not forget that you can use more than one nested system when calculating the trajectory of your service, but it is not recommended to use more than two, or things become very complex. *In Chapter 3*, we posed the following questions that would help with identifying important nested systems to base the equation on; these were:

1. What part of service provision takes up most of my time?

2. What part of service provision is most important?

3. When people think of the service I provide, what is the first task that comes into their minds?

4. What tasks do I do that most typify the nature of the service I provide?

Reviewing your practice with these questions in mind should make nested system recognition fairly straightforward. The thing to remember is that this whole procedure is simply to give you some idea of how your service will develop. It is not intended to provide a comprehensive and faithful reproduction of your service. Remembering this may help you to distinguish what is a nested system and what is 'noise'. If you categorise some of your work as noise, it does not mean that it is any less important, it just means that you have decided to discount it for the purpose of the exercise of Chaos identification.

Stage 3: Dealing with the apples and oranges problem

This means looking at the data you have identified from your holon and your nested systems and finding how to express the amount of time spent on it in a way that shows how nested systems, which seem quite different from each other initially, can be broken down to a common denominator. Once this is done, it should be relatively easy, firstly, to come up with a way of expressing your nested systems as common units (remember we used S/PCU—staff/patient contact units, but it can be anything that works for you) and, secondly, to effectively produce some numerical data that can be plugged in to the population equation. You will need to identify the length of time you spend on nested system activity. You could try keeping a log of your work or asking a colleague to shadow you for a period of time.

Identifying your data set and applying the equation

Once you have identified your nested systems and formatted your data, you should have a seed population and an identified growth parameter. Remember, you should have compared the demands on your system across two time frames; it really does not matter if those time frames are weeks, months, or years, as long as there is little or no population overlap (patients counted in one time frame should not appear in the comparison time frame). With these two pieces of information, you are ready to plug your data into the equation. *Chapter 3* outlines Gleik's (1987) step-by-step method for working the equation out.

Plotting Chaos

When you have taken your calculations as far as you can (or as far as you want to), you can plot them graphically. The graph will help you to decide what the end state of your system is likely to be. You do this by transferring the final population numbers from each iteration of the equation into a separate data sheet and creating a graph from it. If you need to take your calculations to many iterations, you will be better off setting up a spreadsheet programme to run the calculations for you (*Chapter 3* will tell you how). *Chapters 5* and *6* show you the type of end-state graphs you might generate and explains what end state systems they represent.

Play with your parameters

The thing to bear in mind throughout the whole of this exercise is that plotting Chaos is not really predictive; it only shows you what will happen to your system, if all of the original parameters remain unchanged. If you want to see what will happen to your holon if you change growth rate, see less patients, or more, you can use your original seed population data with changed growth rates to do this.

Now, let us examine each of the three attractor states that have been identified in previous chapters and speculate regarding the insights that such material may provide in the context of the forward planning of nursing services. It has been emphasised throughout this work that the data presented is specific to the information presented by the acute pain service described in *Chapter 4*. However, it is possible to examine attractor state outcomes in a broader sense, using the APS2 data as examples rather than rules.

Interpreting the data

Throughout this study, the data utilised has been formatted in staff/patient contact units. In order to make the numerical data generated by the application of the nonlinear equation meaningful, the strategy for interpretation of that data must be understood. In many cases identification of a certain type of attractor state, for example periodic equlibria, might suggest that extra personnel may be the answer. In order to ascertain whether this is an appropriate solution, the S/PCU should be translated back to hours and compared with the contracted hours of the practitioner.

So, if the end point of a system analysis is a fixed point attractor with the population stabilising at a value of 0.600, in order to understand what that means in terms of service provision, we must reverse the formatting that was described in *Chapter 4*. The process is as follows:

$$\textit{Translate } 0.6000 \textit{ in staff / patient contacts} = 6000$$

Remember, we arbitrarily turned our staff/patient contact units into a decimal value because we cannot plug whole numbers into our equation:

$$\textit{Divide } 6000 \textit{ by } 3 \textit{ to calculate working hours} = 2000$$

We divide by 3 because we assumed 3 staff /patient contacts per hour:

$$\textit{Express result as a percentage of } 1687 = 118\%^{*1}$$

1 * 1687 hours equates to 1 WTE within the NHS

This indicates that even though the system is in equilibrium, adequate provision requires more than one pain practitioner to be in place, since the S/PCU translate into 1.18 whole time equivalents (WTE). This reformatting of data can be carried out at any point within the attractor mapping exercise, but is really only of use in those systems, which demonstrate periodic equilibrium, specifically, those likely to become fixed point attractors.

Dynamical systems that demonstrate periodic equilibrium

If the nonlinear data suggests that the end state of a dynamical system is that of a fixed point attractor, the major focus of interest must be upon the length of time it will take to reach that equilibrium state. Once the system is in equilibrium all development will cease, as the entire allocation of practitioner time may be taken up in providing whichever aspects of the service formed the locus of investigation.

If the APS2 data is cited here as an example, it can be seen that, depending upon the 'r' value of the equation the pain service has between 4 and 11 iterations (in the case of the APS2 case study, years) before this equilibrium state is achieved. However, the system is not rigidly conformed. This illustrates the point made by Rapp (1993) regarding stationary systems. It cannot be assumed that the system with a population growth of 20% will survive for four years and then settle into equilibrium because, of course, systems change. The replacement of an experienced pain nurse with a new, inexperienced practitioner might cause a decrease in the 'r' value; extra staff might increase it. This is a very important point that must be emphasised. The application of nonlinear calculations will only predict service development *if all the factors remain unaltered*.

All of this notwithstanding, this type of predictive calculation can inform business planning. A system that will move towards periodic equilibrium is one that will clearly need some extra input, if it is to continue to develop and move forward, rather than simply remain stationary, meeting a need that is already established with no chronological resources to react to future unforeseen needs. In the case of APS2, some consideration must be given to employing at least one extra member of staff within the next four years. This is a simple calculation; by Year 4, the number of staff/patient contacts was shown to be equivalent to 99% of the current pain practitioner's contracted hours. If another full time practitioner is added to the pain service, the workload is halved. In other words, in four years the addition of one member of staff to the acute pain service of APS2 will probably allow the service to function at around the same levels of efficiency as it does today.

In the case of *Figure 5.1*, the system is still a fixed point attractor sliding towards an equilibrium point that will equate to extinction of the service. While this might seem to be opposed to the developments predicted when the 'r' value is raise above 1, it can be argued that the end result is the same—the systems crashes. In the case of what we could term 'positive' growth, such as that shown by *Figures 5.1, 5.2* and *5.3*, it could be argued that the end result was a frozen system, with no possibility of further development. This is equally true of the picture suggested by *Figure 5.1* where there is no growth and no development and, although it has been demonstrated that the slide towards extinction will take a considerable time, it is, nonetheless, inexorable. The action required in a service such as this is to develop strategies that are focussed towards increasing the population that the service deals with. In the case of an acute pain service, this may include the introduction of new pain management techniques, or the amalgamation of acute and chronic pain services to promote viability.

Thus, it can be seen that fixed point attractor systems are amenable to manipulation and this type of forward planning is recommended as good specialist service practice (Pain Society, 1997; Audit Commission, 1998). The semi-predictive nature of the nonlinear data can allow for both budget and service management that will facilitate the continued survival and development of the service in question.

Dynamical systems that are limit cycle attractors

This is the type of system that manifests the 'boom and bust' topology so clearly represented in *Figure 5.4*. Unlike services that are fixed point attractors, a service that is characterised by a limit cycle attractor will never crash or freeze. This is not to say that the service will actually continue to develop. The regular oscillations between two points will translate into a service that has intense periods of activity followed by comparative periods of calm. As can be seen by examining the predicted outcome that was achieved when the growth parameters of the APS2 data were altered to the extent that the system became a limit cycle attractor. This does not necessarily mean that the practitioners will be able to cope. The lower values of the APS2 r = 3 data still equated to more than 100% of a pain practitioner's contracted hours. So, in this instance, the picture is of an already stressed service that has regular periods of increased stress. Therefore, the service manager might be wise to limit population growth until the system becomes a fixed point attractor. This type of system will remain in this pattern forever or until the system parameters change.

An example of this type of limit cycle attractor that many practitioners are familiar with can be seen in the annual winter bed crises. If we view the NHS as a holon and bed occupancy as a nested system, it is obvious that we have a holon which oscillates between manageable patient numbers and overload. Plotted graphically, this would very likely show the typical wave like pattern of a limit cycle attractor.

Chaotic systems

It has been demonstrated that a simple change in the growth parameter of a dynamical pain service system will move the system through various attractor states. This is a practical demonstration of the fundamental chaotic characteristic—sensitive dependence upon initial conditions. It has also been shown that it is possible to predict the end state of a system and to suggest responses that may be palatable to service managers as way of maximising the service potential. However, this changes when a system becomes a chaotic attractor.

Kosko (1994) states that, although Chaos appears to be random, its trajectory can be mapped. This is evidenced by *Figures* 5.5 and 5.6. In terms of service planning and delivery, however, a system that is a chaotic attractor will be one of the most difficult for which to plan. Reference to *Figure 5.6* will show that, not only does the system oscillate between high and low points as was demonstrated by the limit cycle attractor, but that these points are not consistent. In addition, small episodes of stable period cycles, as identified by May (1976), can be seen to make the overall system trajectory appear highly randomised. It is tempting to speculate that such a representation is typical of a service that is characterised by high staff turnover and serious burnout, as staff try to accommodate the haphazard highs and lows of service demand. Consider, the holon would show intense pressure in terms of service delivery which, eventually, would see those staff that can, leaving and being replaced. The new staff might well function efficiently for a period of time (one of May's stable period cycles) but, as pressure and stress mount, the inevitable slide towards Chaos returns.

In terms of service management, this situation is problematical. It cannot be addressed by budgeting for future staff or manipulation of the number of patients who are allowed access to the service. It cannot be addressed by training orientated bank staff to supplement the service in times of predicted stress. The issues might be further augmented by the notion that it will be very difficult for any service that has provided such spectacular growth (in the region of 35–40% per annum) to suddenly limit the number or type of patients that the pain practitioners can see.

In the case of a system that settles into being a chaotic attractor, the only option might be to review the entire service. This could mean entering into negotiation with units that currently access the service, but who do not contribute to the service budget (Pain Society, 1997), or by setting limits to the number of patients that can be seen within one year in an effort to lower the service population growth. The effects of these strategies would be to move the system into being, at least, a limit cycle attractor and, at best, a fixed point attractor. These are not easy decisions to make; any service delivery within the UK health system has, traditionally, often relied upon the altruistic nature of the service providers. However, when a service manager can see how the service will deteriorate, simply by plotting the end state attractor of the chosen nested systems, arguments for change could have more weight.

Summary

The focus of this chapter has been upon the different attractor states that a system might develop. It has been emphasised that the predictive power of the nonlinear analysis is dependent upon all factors remaining the same over the period of iteration. While it is acknowledged that this is unlikely, it has been contended that the purpose of the calculation would be to inform service managers of the potential trajectory of the service provided.

In order to ensure that the data generated by application of the equation to a specific data set is meaningful, it has been demonstrated that S/PCU can be translate into working hours, which can, in turn, be expressed as a percentage of one WTE. Each attractor state has been considered in the light of the potential managerial responses. Although it must be stressed that core population data must be collected on a case by case basis, it is not inappropriate to suggest that, founded upon the APS2 data, any specialist service holon is best assisted by ensuring that the nested systems supporting it are likely to be fixed point attractors. While equilibrium is not necessarily appropriate to any responsive service, it can be argued that the nature of the trajectory does allow for more forward planning in terms of service resources. Limit cycle attractors may offer the same facility, but so much could depend upon the WTE transformation of the low point data. Finally, the manager of any service that demonstrates a predisposition to becoming a chaotic attractor must react swiftly if the whole pain service provision is not to be compromised.

This book has attempted to introduce extremely fundamental concepts of Chaos Theory and to implement them within health service delivery. Without doubt, even the simplest elements of Chaos are difficult to comprehend, but by taking them step-by-step, it is hoped that practitioners will be able to recognise them within every day service delivery. The example used throughout this book has been that of an acute pain service, but the population equation that we have used to detect chaotic outcomes can be usefully employed within any service provision. While the equations and the end state attractor can give a suggestion of possible service outcomes **if all other parameters remain unchanged**, they are indicative rather than predictive. The final chapter of this book will give you the opportunity to practice identifying service direction for yourself. The only real way to grasp the usefulness of Chaos identification is to carry it out for yourself. By doing the exercises that make up *Chapter 7*, you will be able to test your understanding of the principles of the theory before you apply it to your own practice.

7

Trying it out

Well done you have got all the way to the last chapter and you should soon be ready to start looking for the elements of Chaos in your own practice. This chapter is designed to help you to test out the concepts that have been outlined in the rest of the book. Experience has taught me that the best way to get to grips with this particular application of Chaos Theory is to actually put it into practice. This chapter contains three case study type exercises that you can try before you bite the bullet and start to evaluate your own service. Don't worry if the services described are not exactly the same as yours; it is the principles that are important rather than the service to which they are applied. Because different people may approach this exercise in different ways, the example will be followed, not with answers because the answers I arrive at might be different from yours, but by an explanation of how I obtained my results and the approach I took

Exercise 1

The exercise looks at one large service. Look at the holon descriptor to gain a feel for the service you will be examining. For the purpose of this exercise, we will not be identifying a nested system because we will view our holon as if it were a complete system.

Holon

A large teaching hospital situated in a major city with a client population of around 800,000 people. Last year it treated 517,356 patients. For the last five years, the hospital has reported a year on year population growth of between 5% and 1% (mean 3%). If we take the mean 3% growth rate:

a) How do we express our seed population?

b) How do we express our growth rate?

c) What is the end attractor state of the service?

d) How can the service longevity be improved?

Exercise 1: Explanation

a) As you will recall, for this equation to work we have to express the seed population as a number between 0 and 1; therefore our seed population is **0.517356**;

b) This is a tricky one, remember that when we set our population growth rate at 20%, we expressed it as 2.0 when we plugged it into the equation. This holon only has a growth rate of 3% per annum, so we should express it as 0.3;

c)

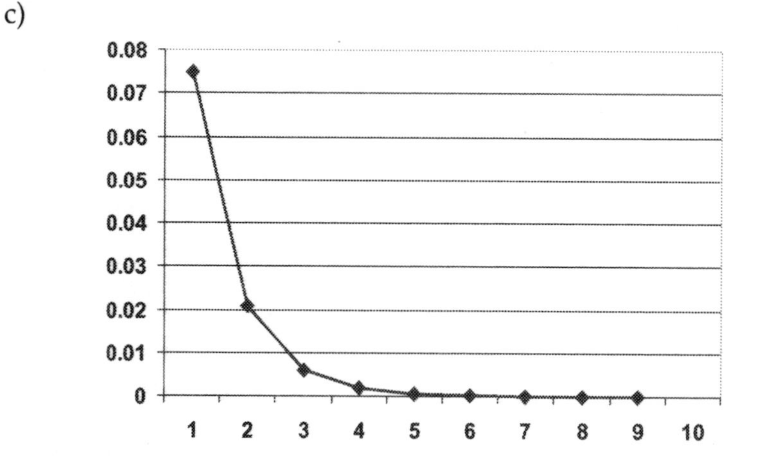

As you can easily see, this system reaches equilibrium very quickly and it becomes a fixed point attractor. With such a small expansion of patient numbers, the service ceases to be viable. When I did this calculation, I set my spreadsheet to record up to 4 decimal places. If I had rounded my calculations up to 3 decimal places, the service would have reached extinction point by Year 5. In reality, this might mean a reduction in financial allocation, department closure or (more likely) amalgamation with another local service and rationalization of resources;

d) In order to survive, it is clear that the service needs to see more patients to increase its patient population growth rate. By playing around with the growth rates on your spreadsheet (and I did it by increasing the growth rate in 0.1 increments) eventually you will get a graph that looks like the following:

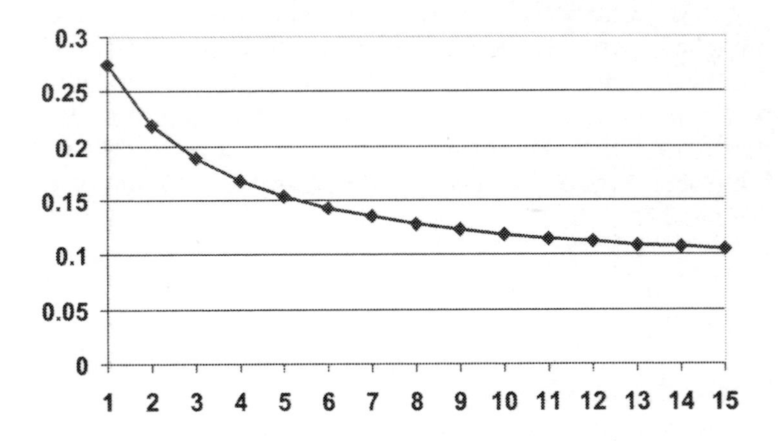

It is plain that the service is still in decline; however, the extinction trajectory is far less steep than our original one. The end state attractor is still a fixed point attractor as the decline towards equilibrium is inexorable, but that decline is greatly slowed. The original graph showed that the holon settled in to its end state by Year 5; on this trajectory, it is still viable at 15 years. I got this result by factoring in a growth rate of 11%. In terms of the holon that I described earlier, this would equate to treating around an extra 41, 000 patients a year, an overall net increase of 8% of the current growth rate.

Exercise 2

This exercise looks at one department within a service, that of an Accident and Emergency (A&E) Department.

Holon

A non-teaching district general hospital. The local client population of 342,138 has a high proportion of elderly people (27.8% of the population). However, as the hospital also services a popular seaside resort, its potential client population is approximately 800,000 over the course of a year.

Nested system

An accident and emergency department. The department has a general accident unit and a specialist dental unit. In the last year, the two units saw a total of 92,847patients, which breaks down into 81,921 patients seen in the general A&E department and 10,926 in the dental A&E . Over the last year, both departments have seen an increase in patients seen of around 17%.

a) Using a growth rate of 17%, plot the trajectory and identify the end state attractor of both departments;

b) Work out what sort of growth rate each department should be aiming for to ensure that the department develops.

Exercise 2: Explanation

a) The graph below shows the end state attractor for the seed population drawn from the general A&E data:.

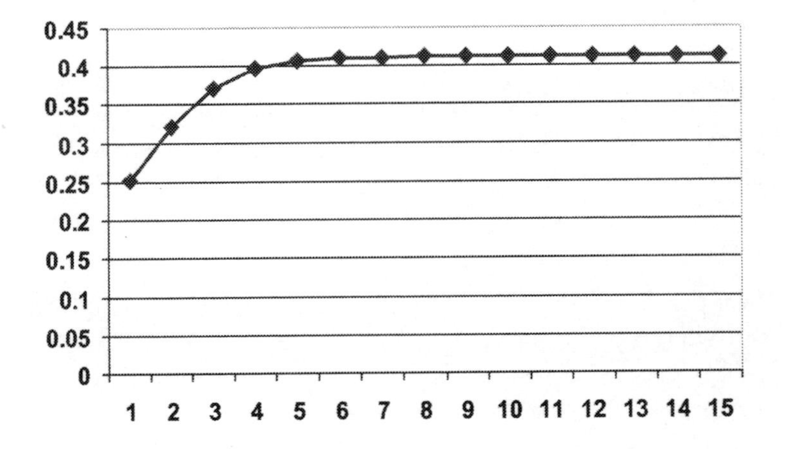

Once again, we see a fixed point attractor state, with a service increase and equilibrium being reached by Year 6. The service is viable, but will need to be reviewed if it is not going to become overloaded. A similar picture is seen when we look at the data from the dental A&E (*Page 92*):

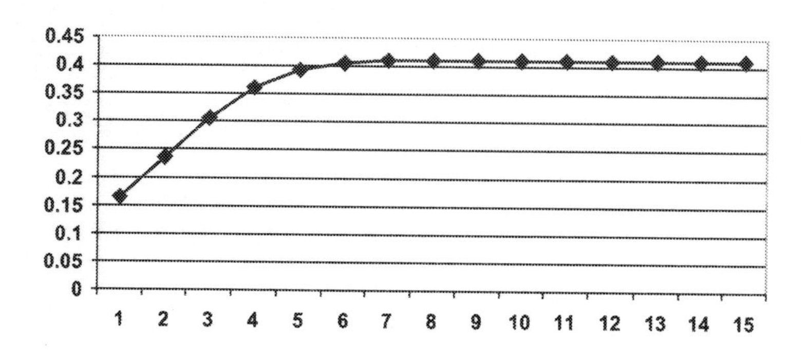

If we put the two services together on one graph we can compare their development:

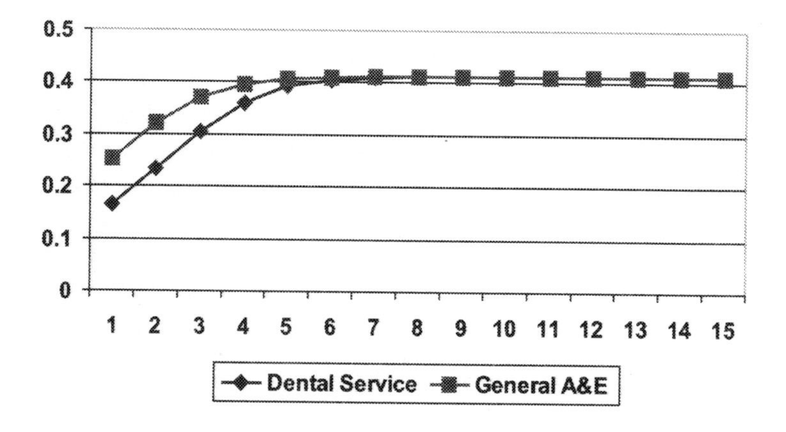

The dental data shows that the service grows quickly to match the work load of the other service, although we can see that it reached equilibrium later, Year 7 rather that Year 5 for the general A&E department.

Exercise 3

This exercise allows you to look at manipulating two related sets of data. Read the outline of the holon and look at the identified nested systems before you start to think about plotting the service.

Holon

A 9–5, five days per week, outpatients radiography department that carries out dual X-ray absorptiometry (DXA) scan to diagnose and manage osteoporosis. Its client group comes from two streams, patients referred by their general practitioner (seed population = 1000 patients) and patients who are involved in clinical trials (seed population = 500 patients).

Nested system

Two nested systems require exploration: the DXA scanning system and production of scan reports. Each scan takes approximately 15 minutes, allowing a through rate of four patients/hour. This figure is the same for both patient streams. Report writing is different for each patient stream. Reports for those patients who have been referred from their GP can be produced at the rate of six/hour, but reports for the research population are produced at the rate of four reports/hour. How can we work out what will happen if:

a) The research population grows by 20%;

or

b) The GP Population grows by 30%?

Think about how you are going to format your data. The questions to ask your self should include:

1. How am I going to express my nested systems?

2. How will I articulate the seed population?

3. How will I format the growth rate?

Exercise 3: Explanation

Let us look at the data for the research patients first. When I looked at the holon, the first thing that was apparent was that the two nested systems were not necessarily incompatible, particularly the systems described for the research group. Report writing and patient scanning for the research group took the same amount of time—four tasks completed per hour. It seemed appropriate to merge that data and express the results as scan/report units (S/RUs), thus:

- 125 hours for scans
- 125 hours for reports
- 250 Scan/Report Units (S/RUs)

This gives us a starting seed population of 250 S/RUs, which we will express as 0.250. We have already noted a growth factor of 20%, which is plugged into the equation as 2. If you have set up your spreadsheet programme try it on that, otherwise use the grid below, I have done the first step for you:

	1 – x	Multiply by x	Multiply by r (r = 2.0)	x (next)
Stage 01: x = 0.250	0.750	0.1875	0.375	0.375
Stage 02: x =				
Stage 03: x =				
Stage 04: x =				
Stage 05: x =				
Stage 06: x =				
Stage 07: x =				
Stage 08: x =				
Stage 09: x =				
Stage 10: x =				

You should have a table that looks something like this:

	1 – x	Multiply by x	Multiply by r (r = 2.0)	x (next)
Stage 01: x = 0.250	0.750	0.1875	0.375	0.375
Stage 02: x = 0.375	0.625	0.234	0.469	0.469
Stage 03: x = 0.531	0.531	0.249	0.498	0.498
Stage 04: x = 0.502	0.502	0.250	0.500	0.500
Stage 05: x = 0.502	0.502	0.250	0.500	0.500
Stage 06: x = 0.502	0.502	0.250	0.500	0.500
Stage 07: x = 0.502	0.502	0.250	0.500	0.500
Stage 08: x = 0.502	0.502	0.250	0.500	0.500
Stage 09: x = 0.502	0.502	0.250	0.500	0.500
Stage 10: x = 0.502	0.502	0.250	0.500	0.500

Does this look like a familiar set of figures? It is very similar to the result we obtained for the APS2 case study in *Chapter 4*. This similarity is more noticeable when we plot the holon trajectory.

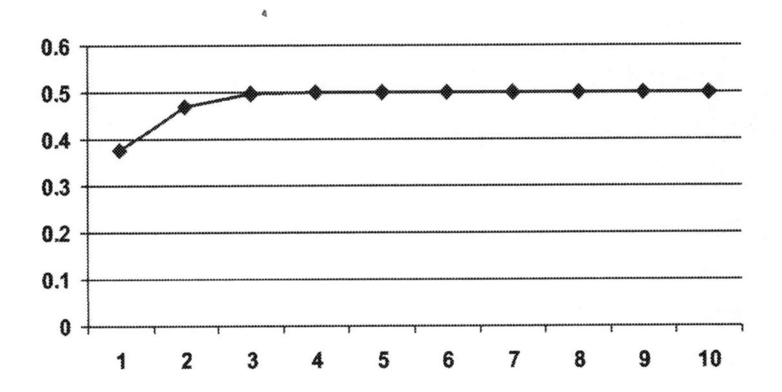

It is clear that the holon will settle in an end state that is equivalent to a fixed point attractor. Remember the point that I made in *Chapter 5*—growth rates generally settle into the same attractor state. The data presented above emphasises that **generally** is not **always**. A growth rate of 30% will generally settle into a limit cycle attractor; however, for this nested system, the end state is that of a fixed cycle attractor. This highlights the importance of always plotting out the population data you develop in order to double check your assumptions.

When we look at the GP patient data, things are slightly different. We still have four patients an hour through the system, but report production is faster for these patients, so our S/RUs are:

- 250 hours for scans
  (1,000 patients:
  4 patients scanned per hour = 1000/4 = 250)
- 166 hours for reports
  (6 reports produced per hour:
  1,000 patients scanned = 1000/6 = 166)

So our seed population is 250 +166 = 416, which we will express as 0.416. The growth rate for GP patients has been set at 30%, so r = 3.0. Fill in the table below:

	1 – x	Multiply by x	Multiply by r (r = 3.0)	x (next)
Stage 01: x =				
Stage 02: x =				
Stage 03: x =				
Stage 04: x =				
Stage 05: x =				
Stage 06: x =				
Stage 07: x =				
Stage 08: x =				
Stage 09: x =				
Stage 10: x =				

When you have done that, plot the holon trajectory to see what the end state will be. You should see something like this:

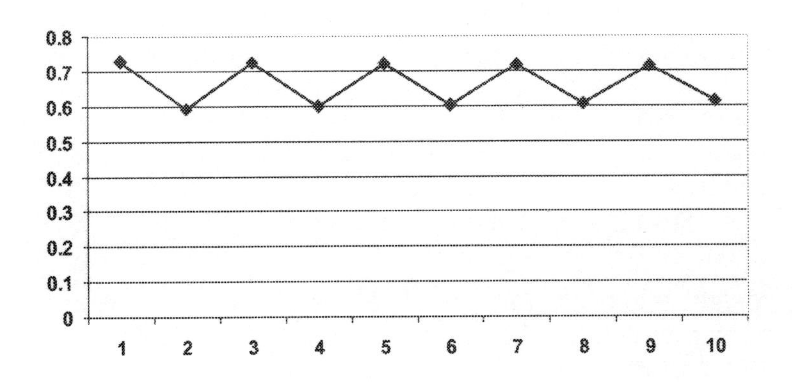

With a growth rate of 30% the holon will become a limit cycle attractor, because your numbers should look like this:

	1 – x	Multiply by x	Multiply by r (r = 3.0)	x (next)
Stage 01: x = 0.416	0.584	0.253	0.729	0.729
Stage 02: x = 0.729	0.271	0.198	0.593	0.593
Stage 03: x = 0.593	0.407	0.241	0.724	0.724
Stage 04: x = 0.724	0.276	0.500	0.5.99	0.599
Stage 05: x = 0.599	0.401	0.240	0.720	0.720
Stage 06: x = 0.720	0.280	0.201	0.604	0.604
Stage 07: x = 0.604	0.396	0.239	0.717	0.717
Stage 08: x = 0.717	0.283	0.208	0.608	0.608
Stage 09: x = 0.608	0.392	0.238	0.715	0.715
Stage 10: x = 0.715	0.285	0.204	0.612	0.612

So what will be the overall effect upon the service, if we take both of our patient populations together and plug them into the equation?

Patient population	1000 + 500/4 = 375
Report production	125 + 166 = 291
S/RU	375 + 291 = 666
Seed population	= 0.666

Ideally, we would amalgamate the growth rates also to give a total growth rate of 50%, but that would take our rate over the 40% barrier, which renders the calculation useless by the introduction of whole numbers, pushing our population beyond the upper limits to growth, so we will compromise by setting the growth rate at the 40% maximum:

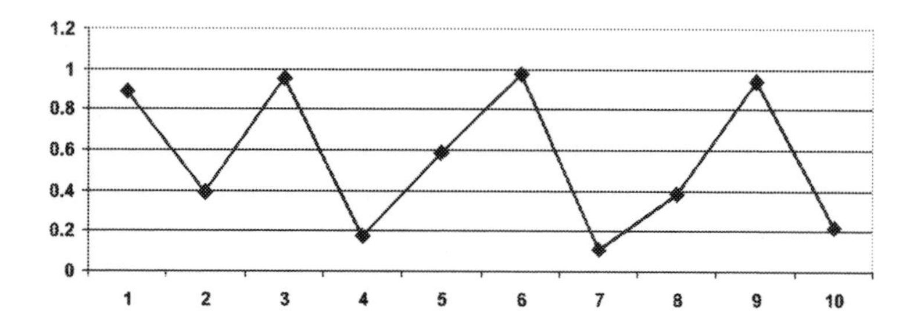

The result is a service in Chaos. Note how the population nears the cut off point of 1 and then plummets down towards extinction as the service falters and develops patient backlog.

If we amalgamated our nested systems, but used 30% as our growth rate our data would look like this:

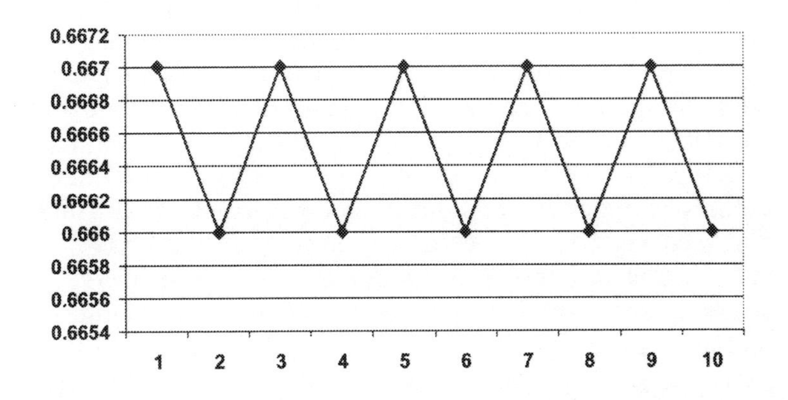

Once again, it is clear that our nested systems will settle into a limit cycle attractor state.

What can we learn from this?

To test that you can apply this data to the management of the service, ask your self the following questions:

1. Can the number of research patients seen by this out patient radiology department be increased?
2. If they are, what will happen to the service?
3. Can the number of GP referrals accepted by this out patient radiology department be increased?
4. If they are, what will happen to the service?
5. Can both elements of the service be increased?
6. If they are, what will happen to the service?
7. If I was running this service what would I do?

Either the research or the GP element of this service can be increased with care. If both elements of the service are increased, the service tilts into Chaos. If the research patient numbers are increased, the service will reach stable equilibrium within four years. If the GP referrals are increased, the service will evolve into a limit cycle attractor and move in and out of high and low patient numbers. If the service is required to see more of both patients, even operating on a 40% growth rather than a 50% one, the service will slip into Chaos. If the population data is amalgamated and a population growth factor of 30% is applied, the service again becomes a limit cycle attractor.

If you had this information as a service manager, you would be able to identify your options. You could decide to increase only one set of patients. You could increase both sets of patients, but by a much smaller amount. You could try to employ extra staff, which would alter the seed population data by altering the number of patients scanned per hour or the number of reports produced. The point is, you could start to think about your options up to four years in advance of system change.

Conclusion

That's it! This book has attempted to take you on a journey exploring the nature and role of Chaos Theory within care delivery. You are now ready to go out and start to look at your own practice to see if you can identify the nested systems within it. Once you have done this, you are well on your way to seeking and finding the elements of chaos within the evolution of the dynamical system that is the care you provide. Remember, any end state you identify will only develop if all other parameters remain unchanged. The purpose of the strategy outlined within this work is not to allow individuals to think, 'Well, we have at least three years before the system crashes', but to facilitate forward planning in order to ensure that a service can continue to provide high quality care for all who need it. It won't give you the answers to all of your problems, but it might, in some small way, help you find the right questions to pose, so that managing those problems gets a bit easier.

References

Audit Commission (1998) *Managing Pain After Surgery. A booklet for nurses*. Audit Commission, London

Barnes-Svarney P (Editorial Director) (1995) *The New York Public Library Science Desk Reference*. Macmillan, New York

Byrne D (1997) Complexity theory and social research. *Social Research Update*. Autumn

Checkland P (1981) *Systems Thinking, Systems Practice*. John Wiley & Sons, Chichester

Checkland P, Scholes J (1990) *Soft Systems Methodology in Action*. John Wiley & Sons, Chichester

Cohen J, Stewart I (1994) *The Collapse of Chaos*. Penguin Books, London

Copnell B (1998) Synthesis in nursing knowledge: an analysis of two approaches. *J Adv Nurs* **27**(4): 870–74

Coppa DF (1993) Chaos Theory suggests a new paradigm for nursing science. *J Adv Nurs* **18**: 985–91

Dawkins R (1989) *The Selfish Gene*, 2nd edn. Oxford University Press, Oxford

Fawcett J (1984) *Analysis and Evaluation of Conceptual Models of Nursing*. FA Davis Company, Philadelphia

Feynman RP (1965) *The Character of Physical Law*. Penguin Edition (1992). Penguin Books, London

Glass L, Young R (1979) Structure and dynamics of neural network oscillators. *Brain Res* **179**: 207–18

Gleick J (1987) *Chaos. Making a New Science*. Penguin Books, London

Gribbin J (2002) *Q is for Quantum. Particle Physics from A to Z*. Phoenix Press, London

Gribbin J (1984) *In Search of Schrödinger's Cat. Quantum Physics and Reality.* Black Swan, London

Griffiths F, Byrne D (1998) General practice and the new science emerging from the theories of 'Chaos and complexity. *Br J Gen Pract* **48:** 1697–99

Haigh C (2000) *Using Chaos Theory as a Predictive Tool in Specialist Pain Services.* MSc Thesis. University of Central Lancashire

Hamilton P, West B, Cherri M, Mackey J, Fisher P (1994) Preliminary evidence of nonlinear dynamics in births to adolescents in Texas, 1964 to 1990. *Theoretic and Applied Chaos in Nursing,* Summer 1

Hawking SW (1987) *A Brief History of Time.* Bantam Press, London

Hayles KN (2000) From Chaos to complexity: Moving through metaphor to practice. *Complexity and Chaos in Nursing,* Vol. 4

Ireson CL (1998) Evaluation of variances in patient outcomes. *Outcome Man Nurs Pract* **2**(4): 162–66

Johnson M (1999a) Guest Editorial: Scholarship, namedropping and the 'Five Minute Test'. *Nurse Educ Today* **19**(8): 599–600

Johnson M (1999b) Observations on positivism and pseudoscience in qualitative nursing research. *J Adv Nurs* **30** (1): 67–73

Johnson M, Long T, White A (2001) Arguments for 'British Pluralism' in qualitative health research. *J Adv Nurs* **33**(2): 243–49

Kosko B (1993) *Fuzzy Thinking. The New Science of Fuzzy Logic.* Harper Collins, London

Kuhn TS (1970) *The Structure of Scientific Revolutions,* 2nd edn. University of Chicago Press, Chicago

Lorenz E (1993) *The Essence of Chaos.* UCL Press, London

Mark BA (1994) Chaos Theory and nursing systems research. *Complexity and Chaos in Nursing.* Vol. 1

Martinerie J, Adam C, Le Van Quyen M *et al* (1998) Epileptic seizures can be anticipated by non-linear analysis. *Nature Med* **4**(10): 1173–76

May RM (1976) Simple mathematical models with very complicated dynamics. *Nature* **261**: 459–67

Oppenhiem AN (1992) *Questionnaire Design, Interviewing and Attitude Measurement*. Pinter Publishers, New York

Pediani R (1996) Chaos and evolution in nursing research *J Adv Nurs* **23**(4): 645–46

Petrillo GA, Glass L (1984) A theory for phase locking of respiration in cats to a mechanical ventilator. *Am J Physiol* **246**: 311–20

Rapp (1993) Chaos in the neurosciences: cautionary tales from the frontier. *Biologist* **40**(2): 89–94

Ray MA (1994) Complex caring dynamics: A unifying model of nursing inquiry. *Complex Chaos Nurs* **1**(1): 23–33

Reed M, Harvey DL (1996) Social Science as the study of complex systems. In: Kiel LD, Elloitt E, eds. *Chaos Theory in Social Sciences.* University of Michigan Press, Ann Arbor: 295–324

Rogers M (1980) Nursing: A science of unitary man. In: Riehl JP, Roy C, eds. *Conceptual Models for Nursing Practice,* 2nd edn. Appleton-Century-Crofts, London: 302–15

Royal College of Surgeons of England, College of Anaesthetists (1990) *Report of the Working Party of Pain After Surgery.* HMSO, London

Sabelli H, Carlson-Sabelli L, Messer J (1994) The process method of comprehensive patient evaluation on the emerging science of complex dynamical systems. *Complex Chaos Nurs* **1**(1): 34–42

Sadar A, Abrams I (1999) *Introducing Chaos.* Icon Books, Cambridge

Sprott JC (1995) *Sprott's Fractal Gallery*: http://sprott.physics.wisc.edu/fractals.htm

Stewart I (1997) *Does God Play Dice? The New Mathematics of Chaos.* Penguin Books, London

Strathern P (1998) *Bohr and Quantum Theory*. Arrow Books, London

The Pain Society (1997) *Provision of Pain Services.* The Association of Anaesthetists, London

Vicenzi A, editorial (1997) Progress in nonlinear dynamical research. *Complex Chaos Nurs* **3**(Summer): 3

Walsh M (2000) Chaos, complexity and nursing. *Nurs Stand* **14**(32): 39–42

Suggested Reading

Byrne D (1997) Complexity theory and social research *Social Research Update,* Autumn

Gleick J (1987) *Chaos. Making a New Science.* Penguin Books, London

Sadar A, Abrams I (1999) *Introducing Chaos.* Icon Books, Cambridge

Sprott JC (1995) *Sprott's Fractal Gallery*

Stewart I (1997) *Does God Play Dice? The New Mathematics of Chaos.* Penguin Books, London

Sweeney K, Griffiths F (2002) *Complexity and Health Care. An Introduction.* Radcliffe Medical Press,

If you want to find out more about Chaos Theory in general, the James Gleick's book is still (in my opinion) the best on the market. Ian Stewart writes in a clear and accessible fashion about the maths underpinning Chaos and to actually see the beauty of Chaos, I highly recommend Sprott's Fractal Gallery, where hundreds of fractals are stored.

Index